Increase Your Puzzle IQ

Also available from Wiley

Increase Your Puzzle IQ

Tips and Tricks for Building Your Logic Power

MARCEL DANESI, PH.D.

John Wiley & Sons, Inc.
New York • Chichester • Weinheim • Brisbane • Singapore • Toronto

This text is printed on acid-free paper.

Published by John Wiley & Sons, Inc.

Library of Congress Cataloging-in-Publication Data:
Danesi, Marcel
Increase your puzzle IQ: tips and tricks for building your logic power / Marcel Danesi
p. cm.
ISBN 978-0-471-15725-0
1. Puzzles. 2. Logic. I. Title.
GV1493.D334 1997
793.73—dc20 96-34965

10 9 8 7 6 5 4 3 2 1

To Alexander, whose gleaming eyes betray an instinctive propensity to search for purpose in the puzzles that life will pose to him as he grows up.

Contents

PREFACE, ix

HOW TO USE THIS BOOK, xiii

ACKNOWLEDGMENTS, xv

1 PUZZLES IN DEDUCTIVE LOGIC, 1
(Puzzles 1–5)

2 PUZZLES IN TRUTH LOGIC, 33
(Puzzles 6–10)

3 PUZZLES IN TRICK LOGIC, 55
(Puzzles 11–24)

4 PUZZLES IN ARITHMETICAL LOGIC, 63
(Puzzles 25–34)

5 PUZZLES IN ALGEBRAIC LOGIC, 81
(Puzzles 35–44)

6 PUZZLES IN COMBINATORY LOGIC, 97
(Puzzles 45–52)

7 PUZZLES IN GEOMETRICAL LOGIC, 121
(Puzzles 53–57)

8 PUZZLES IN CODE LOGIC, 133
(Puzzles 58–66)

9 PUZZLES IN TIME LOGIC, 153
(Puzzles 67–73)

10 PUZZLES IN PARADOX LOGIC, 169
(Puzzles 74–85)

11 A PUZZLE IQ TEST, 181

Preface

People have always been fascinated by conundrums, rebuses, riddles, and enigmas of all kinds. The archeological record makes it clear that there's an innate "puzzle instinct" in our species that has no parallel in any other species. The oldest known cipher—a message laid out in secret code—is a Sumerian text written in cuneiform (i.e., by means of wedge-shaped markings carved in soft clay tablets) that dates back to around 2500 B.C. Puzzles from the Old Babylonian period (1800–1600 B.C.), Egypt (1700–1650 B.C.), and the ancient civilizations of the Orient and the Americas have also been discovered by archeologists.

One of the oldest puzzles known to the Western world is the so-called "Riddle of the Sphinx." In Greek mythology, the Sphinx was a monster with the head and breasts of a woman, the body of a lion, and the wings of a bird. Lying crouched on a rock, she accosted all who were about to enter the city of Thebes by asking them a riddle:

What is it that has four feet in the morning, two at noon, and three at night?

Those who failed to answer the riddle correctly were killed on the spot. On the other hand, if anyone were ever to come up with the correct answer, the Sphinx vowed to destroy herself. When the hero Oedipus solved the riddle by answering, "Man, who crawls on four limbs as a baby [in the morning of life], walks upright on two as an adult [at the noon hour of life], and walks with the aid of a stick in old age [at the twilight of life]," the Sphinx killed herself. For ridding them of this terrible monster, the Thebans made Oedipus their king.

Throughout history, the puzzle instinct has shaped the fancy of many famous personages. Riddle contests were organized by the Biblical Kings Solomon and Hiram. Charlemagne (742–814), the founder of the Holy Roman Empire, Edgar Allan Poe (1809–1849), the great American writer, and Lewis Carroll (1832–1898), who is best known for his two great children's novels, *Alice's Adventures in Wonderland* and *Through the Looking Glass,* devoted countless hours to the making of puzzles. This instinct is alive and well today, as witnessed by the widespread popularity of puzzle magazines, "brain-challenging" sections in newspapers, riddle books for children and adults alike, TV quiz shows, and game tournaments. Millions of people the world over simply seem to enjoy solving puzzles for their own sake. As the great British puzzlist Henry E. Dudeney (1847–1930) aptly put it, it would appear that "Puzzle-solving, like virtue, is its own reward."

Given the popularity and perceived importance of puzzle-solving in societies around the world, it is strange to find that few school courses on puzzle-solving exist, and that most of the puzzle books on the market assume that anyone can solve puzzles without any particular kind of training. There is, of course, a large element of commonsensical thinking involved in solving puzzles. However, it is also true that without a basic understanding of how logical thinking unfolds, and what techniques can be employed to enhance, rehearse, and sustain it, the chances are that the ability to solve puzzles with facility will not emerge in many people. As the great inventor Thomas Edison so aptly remarked: "Genius is 1% intelligence and 99% hard work." Without some form of systematic training and practice in puzzle-solving, frustration, disinterest, and, worst of all, the fear of puzzles will probably ensue. Success at solving puzzles requires that several basic principles and "lines of attack" be grasped firmly and enduringly from the very beginning.

This book is designed precisely to give the beginner in puzzle-solving instruction and training in the basics, and to help more advanced puzzle solvers sharpen their skills. But this book by itself will not guarantee 100% success. It will, however, put you in a better mental position to attack puzzles of any kind more efficiently and intelligently. The kinds of techniques you will be learning and practicing systematically in this book will help you literally "see" why certain puzzles are best approached in particular ways.

Puzzle-solving skills evolve through experience and dynamic interaction with puzzles. Developing the knack of solving puzzles will give you an incomparable feeling of self-confidence. Confidence is more important than any native intellectual ability! Knowing how to solve puzzles in an orderly fashion also has some important hidden by-products. It will prepare you, for instance, to be more competent and successful in taking IQ tests and college entrance examinations.

This book is the result of a course on logic puzzles that I teach at the University of Toronto. Those who take my course are usually self-defined "math phobics" who normally need to acquire puzzle-solving skills to pass intelligence tests, prepare for university entrance exams, become better problem solvers at work, or develop the mental abilities that puzzle-solving entails for some practical purpose. The puzzles included in this book were selected or designed to help such learners gain those skills quickly and enjoyably. I hope you will find the puzzles to be as helpful and pleasant as my students have. It has given me great personal satisfaction over the years to find that those who have worked through the puzzles included in this book, according to the suggested puzzle-solving techniques, have rarely failed to discover how easy and rewarding it is to solve puzzles in logic and mathematics. As the late puzzlist James Fixx once wrote: "Puzzles not only bring us pleasure but also help us to work and learn more effectively."

MARCEL DANESI
University of Toronto, 1996

How to Use This Book

This book contains 105 puzzles spread over ten instructional chapters and a review chapter. There are also two fully explained and solved illustrative examples in each instructional chapter. In total, therefore, you will be exposed to 125 puzzles—which are fully explained and worked out for you to study and grasp. Practice makes perfect. So, even by just reading the solutions to the puzzles provided in this book you will gain a basic knack of what to do and, by the end of the book, acquire a certain adroitness in solving puzzles.

If you are an inveterate puzzle solver, you can use this book as a collection of posers to solve during your leisure hours. You will find plenty here to keep you occupied and entertained. Some of the puzzles are classic nuts; that is, they are puzzles that invariably make their way, in one version or other, into most of the puzzle anthologies available on the market. Others have been created or designed for the specific pedagogical purposes of this book; they have not been selected or constructed in a casual way, but rather they have been carefully composed to teach you how to put your reasoning powers to work in a systematic fashion. No advanced mathematical knowledge is needed to solve any of the puzzles in this book. A brush-up of basic high school geometry and algebra will, however, come in handy to solve the puzzles in a few of the chapters.

After a brief introduction to the puzzle genre to be dealt with in the chapter, each of the first ten chapters is organized as follows:

- How To . . .

  This section is subdivided into three parts: (1) "Puzzle Properties," where the main features of the puzzle genre you will be working with are described briefly; (2) two fully solved examples, with clear guidelines, suggestions, indications, and insights on how to attack or approach the puzzles; (3) a Summary, where the main techniques for solving the puzzles are summarized in point form.

- Puzzles

  The puzzles are numbered consecutively across the chapters. Unless you are already an experienced puzzle solver, you should study the illustrative examples carefully and attentively, trying very hard to familiarize yourself with, and even memorize, the techniques suggested, before attempting to solve the puzzles on your own. However, if you are using this book for recreation, or for practice in puzzle-solving, you can certainly skip the examples.

- ANSWERS AND SOLUTIONS

  Answers and step-by-step solutions to the puzzles are given at the end of the chapter. You should not read these until you have attempted to solve the puzzles on your own first, no matter how frustrated you might be with any particular puzzle. After a while, you will get the knack of how to go about attacking puzzles of all kinds. If you get the answer using a different line of attack than the one suggested in the book, you should still study the given solution simply to get a different perspective on the puzzle-solving process itself. This, too, will enhance your puzzle-solving skills. Incidentally, these skills are not correlated with quickness of thought. A slow thinker can solve a puzzle just as successfully as a fast one can.

Some chapters are longer than others. This has been necessitated by the "mechanics" required to solve certain types of puzzles. This does not in any way mean that they are more *difficult* to solve—just more *elaborate.* Incidentally, I have avoided labeling certain puzzles as more difficult than others, as is the practice in some puzzle anthologies, because such a grading system does not mean very much in my opinion. In effect, all the puzzles in this book are hard to solve, especially if you have not solved puzzles before, or have been frustrated by puzzle-solving in the past. The primary purpose of this book is to make puzzle-solving easier. Once you have grasped the underlying principles of how to solve specific kinds of puzzles, and gained facility by the actual experience of solving puzzles, you will have acquired what it takes to solve all kinds of puzzles in the future.

Acknowledgments

It is impossible to thank all those who have given me feedback, advice, and encouragement during the writing of this book. I am especially grateful to my good friend and colleague, Professor Ed Barbeau of the Department of Mathematics (University of Toronto), for all the insights on the nature of puzzles and problem-solving with which he has provided me over the years. I also want to express my gratitude to Judith McCarthy of John Wiley for her kindly interest in my work and for having called it to the attention of the publisher. And I must also mention that it has been a pleasure working with Chris Jackson, who edited the manuscript and whose many excellent suggestions have allowed me to improve it considerably. I thank you all from the bottom of my heart!

1 Puzzles in Deductive Logic

Developing the ability to think logically is a prerequisite for solving puzzles in mathematics, science, or life, for that matter. The great English puzzlist Henry E. Dudeney (1847–1930) went so far as to claim that "the history of puzzles entails nothing short of the actual story of the beginnings and development of exact thinking in man." Puzzlists assume that the solvers of their brainteasers already possess a rudimentary knowledge of the "laws" or "rules" of logic. They are called rules because they form a system of thinking that is not unlike the grammar of a language. As a matter of fact, the word *logic* comes from the ancient Greek noun *logos,* which means "word, speech, thought."

In logic, the basic form of reasoning by which a specific conclusion is inferred from one or more premises is called *deduction.* In valid deductive reasoning, the conclusion must be true if all the premises are true. Thus, if it is agreed that all human beings have one head and two arms, and that Beatrice is a human being, then it can logically be concluded that Beatrice has one head and two arms. This is an example of deductive reasoning, an argument in which two premises are given and a logical conclusion is deduced from them.

It is therefore logical to start off with puzzles that are based purely on deduction and inference. These are designed to help you stretch your mental muscles.

How To . . .

Puzzles in deduction require the ability to think clearly; they involve no play on words, no guessing, and no technical know-how. Only commonsense knowledge applies here: for example, that a mother is older than her children, that an only child has no brothers or sisters, and so on.

Puzzle Properties

Before solving puzzles in logical deduction, you will first have to recognize them as such. Here are some general characteristics to look for that will help you distinguish these puzzles from others:

- A puzzle in logical deduction will invariably ask you to figure out how two or more sets of facts relate to each other: for example, what first name belongs with which last name, what jobs certain persons do, and so on.
- The conditions stated in the puzzle require that specific connections be made. Most puzzles will start with a statement such as this one: *John, Mary, and Donna hold B.A., M.A., and Ph.D. degrees from a reputable university, but not necessarily in that order. . . .* The puzzle will then give you other bits of information so that you can establish which degree, the *B.A.*, the *M.A.*, or the *Ph.D.*, each person, *John, Mary,* and *Donna,* holds, since one person holds only one degree.
- Occasionally, a puzzle might entail some arithmetical calculations. Such calculations are not ends in themselves; they are simply facts to be employed in solving the puzzle.

Example 1 The following is a version of a classic puzzle that is included in most collections of deduction puzzles.

> In a certain company, the positions of director, engineer, and accountant are held by Bob, Janet, and Shirley, but not necessarily in that order. The accountant, who is an only child, earns the least. Shirley, who is married to Bob's brother, earns more than the engineer. What position does each person fill?

Puzzles such as this one can boggle the mind, unless you attack them in some systematic and orderly fashion. The first thing to do, therefore, is to draw a *cell chart,* putting the positions—director, engineer, accountant—on one axis and the names of the persons—Bob, Janet, Shirley—on the other:

	DIRECTOR	ENGINEER	ACCOUNTANT
BOB			
JANET			
SHIRLEY			

← *Positions*

↑ *Persons*

Putting X's and ●'s in this simple chart will allow you to keep track visually of the facts as you discover them:

- If you conclude, for instance, that one of the three people cannot be the director, then place an X in the cell opposite his or her name under the column headed director.
- If you deduce that one of the three is the engineer, then put a ● in the cell opposite his or her name under the column headed engineer, eliminating the remaining cells in the column (since there can be only one engineer) with X's.

- The solution is complete when you have placed exactly one ● in each row/column successfully.

You are told that: (1) the accountant is an only child, and (2) Bob has a brother (to whom, incidentally, Shirley is married). So, clearly, you can eliminate Bob as the accountant, placing an × in the cell opposite his name under the column headed accountant:

	DIRECTOR	ENGINEER	ACCOUNTANT
BOB			×
JANET			
SHIRLEY			

You are also told that: (1) the accountant earns the least of the three, and (2) Shirley earns more than the engineer. From these facts, two obvious things about Shirley can be established: (1) she is not the accountant (who earns the least, while she earns more than someone else); (2) she is not the engineer (for she earns more than he or she does). To keep track of these two facts, enter two ×'s in their appropriate cells, eliminating accountant and engineer as possibilities for Shirley:

	DIRECTOR	ENGINEER	ACCOUNTANT
BOB			×
JANET			
SHIRLEY		×	×

Look closely at the chart. Do you see that the only cell left under accountant is opposite Janet? Therefore, by the process of elimination, Janet is the accountant. Show this by putting a ● opposite her name in the cell, and eliminating all other possibilities for Janet with ×'s, because Janet can hold only one of the stated positions—if she is the accountant, then, logically, she is neither the director nor the engineer:

	DIRECTOR	ENGINEER	ACCOUNTANT
BOB			×
JANET	×	×	●
SHIRLEY		×	×

Once again, look at the chart. Do you see that the only cell left under engineer is opposite Bob? So, put a ● in the cell opposite Bob under engineer, eliminating all other possibilities with ×'s:

	DIRECTOR	ENGINEER	ACCOUNTANT
BOB	×	●	×
JANET	×	×	●
SHIRLEY		×	×

Look at the chart one last time. Do you see that the only cell left opposite Shirley is under director?

	DIRECTOR	ENGINEER	ACCOUNTANT
BOB	×	●	×
JANET	×	×	●
SHIRLEY	●	×	×

The solution is now complete. With the aid of the cell chart technique, it has been a rather straightforward task to establish that Bob is the engineer, Janet the accountant, and Shirley the director. The chart helped you do two important things: (1) keep track of the connections you made, and (2) show you further connections *on its own* as you went along.

Example 2 In the puzzle above, the cell chart allowed you to keep track of and/or make connections between two sets of facts (names and positions). But what if the puzzle requires you to correlate more than two sets? Can you still use a cell chart?

Consider the following example, which involves correlating three sets of data.

Anna, Cindy, Nancy, Rose, and Sonia—one of whose last name is Mill—were recently hired as concession clerks at a large cinemaplex. Each woman sells only one kind of fare. From the following clues, determine each woman's full name and the type of refreshment she sells.

1. Rose, whose last name is not Wilson, does not sell popcorn.
2. The Dunne woman does not sell candy or soda.
3. The five women are Nancy, Rose, the Smith woman, the Carter woman, and the woman who sells ice cream.
4. Anna's last name is neither Wilson nor Carter. Neither Anna nor Carter is the woman who sells candy.
5. Neither the peanut vendor nor the ice cream vendor is named Sonia or Dunne.

The three sets of facts that you will have to connect to each other are: (1) the first names of the women (Anna, Cindy, Nancy, Rose, Sonia); (2) their last names (Carter, Dunne, Mill, Smith, Wilson); and (3) the refreshments they sell (candy, ice cream, peanuts, popcorn, soda). To correlate these sets in a cell chart, it will be necessary to repeat one of them, say the refreshments set, to the right of and under, say, the last names set. This makes it possible to register correlations with X's and ●'s among refreshments, first names, and last names simultaneously:

		Last names					Refreshments				
		Carter	Dunne	Mill	Smith	Wilson	Candy	Ice Cream	Peanuts	Popcorn	Soda
Names	Anna										
	Cindy										
	Nancy										
	Rose										
	Sonia										
Refreshments	Candy										
	Ice Cream										
	Peanuts										
	Popcorn										
	Soda										

← *Repeated Set* ↑

Clue 1 tells you two things about Rose: namely, (1) that she is not Ms. Wilson; and (2) that she does not sell popcorn. Register these facts by putting one X in the cell under Wilson opposite Rose, and another X in the cell under popcorn in the right-hand refreshments set also opposite Rose:

		Last names					Refreshments				
		Carter	Dunne	Mill	Smith	Wilson	Candy	Ice Cream	Peanuts	Popcorn	Soda
Names	Anna										
	Cindy										
	Nancy										
	Rose					×				×	
	Sonia										
Refreshments	Candy										
	Ice Cream										
	Peanuts										
	Popcorn										
	Soda										

The second clue tells you that Ms. Dunne does not sell candy or soda. This means, of course, that you can now put X's in the cells opposite candy and soda in the lower refreshments set under Dunne:

		Last names					Refreshments				
		Carter	Dunne	Mill	Smith	Wilson	Candy	Ice Cream	Peanuts	Popcorn	Soda
Names	Anna										
	Cindy										
	Nancy										
	Rose					×				×	
	Sonia										
Refreshments	Candy		×								
	Ice Cream										
	Peanuts										
	Popcorn										
	Soda		×								

Clue 3 identifies the five women individually as: Nancy, Rose, Ms. Smith, Ms. Carter, and the ice cream vendor. From this, you can establish logically that: (1) Nancy and Rose are neither Ms. Carter nor Ms. Smith, because these are *four different women* (a woman named Nancy, another woman named Rose, a third woman named Ms. Carter, and a fourth woman named Ms. Smith); and (2) that Nancy, Rose, Ms. Carter, and Ms. Smith do not sell ice cream, again because the clue lists them as different women (a woman named Nancy, a woman named Rose, a woman named Ms. Smith, a woman named Ms. Carter, and a woman who sells ice cream). So, you can now put ×'s in the chart as follows: (1) in the cells under Carter and Smith opposite both Nancy and Rose; (2) in the cells under ice cream opposite both Nancy and Rose in the right-hand refreshments set; and (3) in the cells opposite ice cream under Carter and Smith in the lower refreshments set. The chart will then look like this:

		Last names					Refreshments				
		Carter	Dunne	Mill	Smith	Wilson	Candy	Ice Cream	Peanuts	Popcorn	Soda
Names	Anna										
	Cindy										
	Nancy	×			×			×			
	Rose	×			×	×		×		×	
	Sonia										
Refreshments	Candy		×								
	Ice Cream	×			×						
	Peanuts										
	Popcorn										
	Soda		×								

Clue 4 tells you that Anna is neither Ms. Wilson nor Ms. Carter. It also tells you that neither Anna nor Ms. Carter sells candy. This new information allows you to put ×'s in the chart as follows: (1) in the cells under Carter and Wilson opposite Anna; (2) in the cell opposite candy under Carter in the lower refreshments set; and (3) in the cell under candy opposite Anna in the right-hand refreshments set. The chart will then look like this:

		Last Names					Refreshments				
		Carter	Dunne	Mill	Smith	Wilson	Candy	Ice Cream	Peanuts	Popcorn	Soda
Names	Anna	×				×	×				
	Cindy										
	Nancy	×			×			×			
	Rose	×			×	×		×		×	
	Sonia										
Refreshments	Candy	×	×								
	Ice Cream	×			×						
	Peanuts										
	Popcorn										
	Soda		×								

Clue 5 tells you that neither Sonia nor Ms. Dunne sells peanuts or ice cream. From this, you can deduce that: (1) Sonia is not Ms. Dunne, showing this with an × in the cell under Dunne opposite Sonia; (2) Sonia does not sell peanuts or ice cream, showing this with ×'s in the cells under peanuts and ice cream opposite Sonia in the right-hand refreshments set; and (3) Ms. Dunne does not sell peanuts or ice cream, showing this with ×'s in the cells opposite peanuts and ice cream under Dunne in the lower refreshments set:

		Last Names					Refreshments				
		Carter	Dunne	Mill	Smith	Wilson	Candy	Ice Cream	Peanuts	Popcorn	Soda
Names	Anna	×				×	×				
	Cindy										
	Nancy	×			×			×			
	Rose	×			×	×		×		×	
	Sonia		×					×	×		
Refreshments	Candy	×	×								
	Ice Cream	×	×		×						
	Peanuts		×								
	Popcorn										
	Soda		×								

Now, look at the lower refreshments set. Do you see that there is only one cell left opposite the popcorn vendor—namely, under Dunne? Go ahead and put a ● in that cell. Ms. Dunne is, therefore, the popcorn vendor. Logically, no other last name can be connected to the popcorn vendor. So, eliminate all other possibilities with ×'s opposite popcorn vendor.

		Last names					Refreshments				
		Carter	Dunne	Mill	Smith	Wilson	Candy	Ice Cream	Peanuts	Popcorn	Soda
Names	Anna	×				×	×				
	Cindy										
	Nancy	×			×			×			
	Rose	×			×	×		×		×	
	Sonia		×					×	×		
Refreshments	Candy	×	×								
	Ice Cream	×	×		×						
	Peanuts		×								
	Popcorn	×	●	×	×	×					
	Soda		×								

With no other clues, at this point we seem to have reached an impasse. But recall from example 1 above that the chart *by itself* can perhaps help you go further, because it might reveal connections to be made *on its own*. Let's see.

- Since Sonia is not Ms. Dunne, as you can see from the × in the cell under Dunne opposite Sonia, she is not, therefore, the popcorn vendor (who is Ms. Dunne, as you have just discovered). Go ahead and register this by putting an × in the cell under popcorn opposite Sonia in the right-hand refreshments set.
- Since Rose is not the popcorn vendor, as you can see from the × in the cell opposite Rose under popcorn in the right-hand refreshments set, she is not, therefore, Ms. Dunne (who is the popcorn vendor, as you know). So, put an × in the cell under Dunne opposite Rose:

		Last names					Refreshments				
		Carter	Dunne	Mill	Smith	Wilson	Candy	Ice Cream	Peanuts	Popcorn	Soda
Names	Anna	×				×	×				
	Cindy										
	Nancy	×			×			×			
	Rose	×	×		×	×		×		×	
	Sonia		×					×	×	×	
Refreshments	Candy	×	×								
	Ice Cream	×	×		×						
	Peanuts		×								
	Popcorn	×	●	×	×	×					
	Soda		×								

Now, do you see that there is only one cell left opposite Rose for her last name—under Mill? Rose is, therefore, Ms. Mill. Show this by putting a ● in that cell and eliminating Mill as a possibility for the other first names with ×'s:

		LAST NAMES					REFRESHMENTS				
		CARTER	DUNNE	MILL	SMITH	WILSON	CANDY	ICE CREAM	PEANUTS	POPCORN	SODA
NAMES	ANNA	×		×		×	×				
	CINDY			×							
	NANCY	×		×	×			×			
	ROSE	×	×	●	×	×		×		×	
	SONIA		×	×				×	×	×	
REFRESHMENTS	CANDY	×	×								
	ICE CREAM	×	×		×						
	PEANUTS		×								
	POPCORN	×	●	×	×	×					
	SODA		×								

Now, note that Rose does not sell ice cream, as you can see from the × in the cell under ice cream in the right-hand refreshments set opposite Rose. Since you have just established that Rose is Ms. Mill, then logically Ms. Mill does not sell ice cream. Show this by putting an × in the cell opposite ice cream under Mill in the lower refreshments set:

		LAST NAMES					REFRESHMENTS				
		CARTER	DUNNE	MILL	SMITH	WILSON	CANDY	ICE CREAM	PEANUTS	POPCORN	SODA
NAMES	ANNA	×		×		×	×				
	CINDY			×							
	NANCY	×		×	×			×			
	ROSE	×	×	●	×	×		×		×	
	SONIA		×	×				×	×	×	
REFRESHMENTS	CANDY	×	×								
	ICE CREAM	×	×	×	×						
	PEANUTS		×								
	POPCORN	×	●	×	×	×					
	SODA		×								

Look at the lower refreshments set and you will see that there is only one cell left opposite the ice cream vendor—namely, under Wilson. Show this with a ●, eliminating the other possibilities with ×'s:

		Last names					Refreshments				
		Carter	Dunne	Mill	Smith	Wilson	Candy	Ice Cream	Peanuts	Popcorn	Soda
Names	Anna	×		×		×	×				
	Cindy			×							
	Nancy	×		×	×			×			
	Rose	×	×	●	×	×		×		×	
	Sonia		×	×				×	×	×	
Refreshments	Candy	×	×			×					
	Ice Cream	×	×	×	×	●					
	Peanuts		×			×					
	Popcorn	×	●	×	×	×					
	Soda		×			×					

Now, look at the right-hand refreshments set. There you will see that Nancy, Rose, and Sonia do not sell ice cream, because there are ×'s in the cells opposite their names under ice cream. So, not one of them is Ms. Wilson (who is the ice cream vendor, as you have just discovered). Show this by putting ×'s in the cells under Wilson opposite Nancy and Sonia (there is one there already in the cell opposite Rose):

		Last names					Refreshments				
		Carter	Dunne	Mill	Smith	Wilson	Candy	Ice Cream	Peanuts	Popcorn	Soda
Names	Anna	×		×		×	×				
	Cindy			×							
	Nancy	×		×	×	×		×			
	Rose	×	×	●	×	×		×		×	
	Sonia		×	×		×		×	×	×	
Refreshments	Candy	×	×			×					
	Ice Cream	×	×	×	×	●					
	Peanuts		×			×					
	Popcorn	×	●	×	×	×					
	Soda		×			×					

The chart now shows simultaneously that: (1) Cindy is Ms. Wilson, because the only cell left under Wilson is opposite Cindy; and (2) Nancy is Ms. Dunne, because the only cell left opposite Nancy is under Dunne. So, put ●'s in their appropriate cells to show these two facts and ×'s to eliminate the other possibilities (as you have been doing above). The chart then looks like this:

		Last Names					Refreshments				
		Carter	Dunne	Mill	Smith	Wilson	Candy	Ice Cream	Peanuts	Popcorn	Soda
Names	Anna	×	×	×		×	×				
	Cindy	×	×	×	×	●					
	Nancy	×	●	×	×	×		×			
	Rose	×	×	●	×	×		×		×	
	Sonia		×	×		×		×	×	×	
Refreshments	Candy	×	×			×					
	Ice Cream	×	×	×	×	●					
	Peanuts		×			×					
	Popcorn	×	●	×	×	×					
	Soda		×			×					

The chart now reveals further that: (1) Anna is Ms. Smith, because the only cell left opposite Anna is under Smith; and (2) Sonia is Ms. Carter, because the only cell left under Carter is opposite Sonia. Put the two ●'s and the × in the last names set in their correct cells. This completes the last names set:

		Last Names					Refreshments				
		Carter	Dunne	Mill	Smith	Wilson	Candy	Ice Cream	Peanuts	Popcorn	Soda
Names	Anna	×	×	×	●	×	×				
	Cindy	×	×	×	×	●					
	Nancy	×	●	×	×	×		×			
	Rose	×	×	●	×	×		×		×	
	Sonia	●	×	×	×	×		×	×	×	
Refreshments	Candy	×	×			×					
	Ice Cream	×	×	×	×	●					
	Peanuts		×			×					
	Popcorn	×	●	×	×	×					
	Soda		×			×					

Now, note in the lower refreshments set that Ms. Dunne is the popcorn vendor, as you can see by the fact that there is a ● in the cell under Dunne opposite popcorn. Since Nancy is Ms. Dunne, then Nancy is the popcorn vendor. Show this in the usual fashion in the right-hand refreshments set. Note also in the lower refreshments set that Ms. Wilson is the ice cream vendor, as you can see by the fact that there is a ● in the cell under Wilson opposite ice cream. Since you have established that Cindy is Ms. Wilson, then logically Cindy is the ice cream vendor. Show this as well in the usual fashion in the right-hand refreshments set:

		Last names					Refreshments				
		Carter	Dunne	Mill	Smith	Wilson	Candy	Ice Cream	Peanuts	Popcorn	Soda
Names	Anna	×	×	×	●	×	×	×		×	
	Cindy	×	×	×	×	●	×	●	×	×	×
	Nancy	×	●	×	×	×	×	×	×	●	×
	Rose	×	×	●	×	×		×		×	
	Sonia	●	×	×	×	×		×	×	×	
Refreshments	Candy	×	×			×					
	Ice Cream	×	×	×	×	●					
	Peanuts		×			×					
	Popcorn	×	●	×	×	×					
	Soda		×			×					

Now, focus your attention on the right-hand refreshments set. Note that Anna does not sell candy, as you can see by the fact that there is an × in the cell opposite Anna under candy. Since you have established that Anna is Ms. Smith, then logically Ms. Smith does not sell candy. Show this with an × in the appropriate cell in the lower refreshments set. Now, the only cell left opposite candy in the lower refreshments set is under Mill. Show this by putting a ● in that cell (adding ×'s to eliminate the other possibilities). At the same time, you can show that Rose, who is Ms. Mill, is the candy vendor in the right-hand refreshments set with a ● in the appropriate cell, adding ×'s as required:

		Last names					Refreshments				
		Carter	Dunne	Mill	Smith	Wilson	Candy	Ice Cream	Peanuts	Popcorn	Soda
Names	Anna	×	×	×	●	×	×	×		×	
	Cindy	×	×	×	×	●	×	●	×	×	×
	Nancy	×	●	×	×	×	×	×	×	●	×
	Rose	×	×	●	×	×	●	×	×	×	×
	Sonia	●	×	×	×	×	×	×	×	×	
Refreshments	Candy	×	×	●	×	×					
	Ice Cream	×	×	×	×	●					
	Peanuts		×	×		×					
	Popcorn	×	●	×	×	×					
	Soda		×	×		×					

Look at the right-hand refreshments set. Do you see that there is only one cell left opposite Sonia—namely, under soda? Putting the appropriate ● and × in the right-hand refreshments set, you will then see that the only cell left under peanuts in that set is opposite Anna:

		Last names					Refreshments				
		Carter	Dunne	Mill	Smith	Wilson	Candy	Ice Cream	Peanuts	Popcorn	Soda
Names	Anna	×	×	×	●	×	×	×	●	×	×
	Cindy	×	×	×	×	●	×	●	×	×	×
	Nancy	×	●	×	×	×	×	×	×	●	×
	Rose	×	×	●	×	Ice	●	×	×	×	×
	Sonia	●	×	×	×	×	×	×	×	×	●
Refreshments	Candy	×	×	●	×	×					
	Ice Cream	×	×	×	×	●					
	Peanuts		×	×		×					
	Popcorn	×	●	×	×	×					
	Soda		×	×		×					

Now that you know that Anna is the peanut vendor and that Anna is Ms. Smith, you can establish that Ms. Smith is the peanut vendor. Show this by putting a ● in the cell under Smith opposite peanuts in the lower refreshment set (adding ×'s in their appropriate cells). You will then see that the only cell left opposite soda in the lower refreshments set is under Carter:

		Last names					Refreshments				
		Carter	Dunne	Mill	Smith	Wilson	Candy	Ice Cream	Peanuts	Popcorn	Soda
Names	Anna	×	×	×	●	×	×	×	●	×	×
	Cindy	×	×	×	×	●	×	●	×	×	×
	Nancy	×	●	×	×	×	×	×	×	●	×
	Rose	×	×	●	×	×	●	×	×	×	×
	Sonia	●	×	×	×	×	×	×	×	×	●
Refreshments	Candy	×	×	●	×	×					
	Ice Cream	×	×	×	×	●					
	Peanuts	×	×	×	●	×					
	Popcorn	×	●	×	×	×					
	Soda	●	×	×	×	×					

The chart is now complete. You have established that Anna Smith is the peanut vendor, Cindy Wilson the ice cream vendor, Nancy Dunne the popcorn vendor, Rose Mill the candy vendor, and Sonia Carter the soda vendor.

□□■ Summary

Although it is not always possible to do so, most puzzles in logical deduction can be solved with the aid of cell charts like the ones used in the illustrative examples above. Such charts allow you to keep track of facts and to make connections as you go along. Even when it is not possible to construct such charts for certain deduction puzzles, it is always useful to jot down and relate facts to each other in some systematic and orderly fashion.

In summary, when solving most puzzles in logical deduction, keep the following things in mind:

- Check to see, by a quick reading of the puzzle, how many sets of facts are to be included in the cell chart.
- If there are two sets to be correlated to each other, set up the chart so as to display the two given sets along two axes (example 1).
- If there are more than two sets, then repeat the additional set(s) to the *right* and *under* a selected set (or sets) (example 2).
- Put an ✕ in a cell to eliminate a connection between two given facts, and a ● to establish a connection between two given facts.
- Look the chart over from time to time, because the chart *on its own* might show you what further connections can be made.
- Every once in a while, a trial-and-error line of reasoning might be required: that is, you might have to try out one, two, or more plausible hypotheses. The one that yields no contradictions is the correct one.

Puzzles 1–5

Answers, along with step-by-step solutions, can be found at the end of the chapter.

1 In the Smith family, the father, the mother, and their 5-year-old daughter are called by their nicknames—Biff, Jiff, and Spiff—but not necessarily in that order. Jiff is not the mother, and Biff is not the daughter. Jiff is older than Biff. How are the three related to each other?

2 Mr. Chow, Mr. Dow, and Mr. Frow work as carpenter, painter, and plasterer, but not necessarily in that order. The plasterer makes more money than the painter. Mr. Frow makes more money than Mr. Dow. The painter and the carpenter always go out together with their wives on weekends. The carpenter and the plasterer have known each other since childhood. Mr. Frow has never met Mr. Dow. What is each man's occupation?

3 Last week, three boys and three girls made dates to go to the prom. One girl was dressed in red, one in green, and one in blue. The boys also wore outfits in the same three colors. While the three couples were dancing, the boy in red said to the girl in green and to her partner: *"Not one of us is dancing with a partner dressed in the same color."* Can you tell who went with whom to the prom?

4 Five women and one man, named Peter, including a geologist, were recently invited as experts to an international conference held at the United Nations on the state of the environment. Can you determine who is engaged in each profession?

1. Karen debated Lori and the meteorologist at the beginning of the conference.
2. Peter is not the physicist.
3. Mary is not the urban planner.
4. Joan is neither the meteorologist nor the biologist.
5. At the end of the conference, the six experts had a general discussion around a table. The debaters were: the physicist, Karen, Joan, the zoologist, the female urban planner, and Paula.

5 Five children, aged three, four, five, six, and seven, go to the same judo class each Saturday afternoon. Can you determine the full names and ages of the children?

1. Every Saturday Mrs. Grant goes to work, so she leaves her children with Mrs. Winn to take to the class with her. Mrs. Winn's daughter is younger than Mrs. Grant's children.
2. Tam is older than Larry but younger than the Guild child.
3. The Brown girl is 2 years older than Lori.
4. Rita's mother, who is sometimes home on Saturday, occasionally takes Gary to the class, while his mother goes shopping. However, she never takes either Grant child.

Note that Tam, Lori, and Rita are the names of girls and that Larry and Gary are the names of boys.

Answers and Solutions

1 **Answer:** *Biff is the mother, Jiff the father, and Spiff the daughter.*

Solution: To correlate the two sets of facts given in the puzzle—nicknames (Biff, Jiff, Spiff) and family relations (father, mother, daughter)—set up a simple two-axis chart, similar to the one used in example 1 above:

	FATHER	MOTHER	DAUGHTER
BIFF			
JIFF			
SPIFF			

Place an X in the cell opposite a name under a family relation when you discover that the name is to be eliminated as a possibility for that family member. Put a ● opposite a name under a family relation when you establish that the name is, in fact, the nickname of that family member, eliminating the remaining cells in the respective row and column with X's.

You are told that Jiff is not the mother. So, eliminate the mother possibility opposite Jiff with an ×:

	FATHER	MOTHER	DAUGHTER
BIFF			
JIFF		×	
SPIFF			

You are also told that Biff is not the daughter. So, you can eliminate daughter as a possibility for Biff:

	FATHER	MOTHER	DAUGHTER
BIFF			×
JIFF		×	
SPIFF			

Now, consider the statement *Jiff is older than Biff.* Since the daughter is the youngest member of the family, you can safely conclude that Jiff is not the daughter:

	FATHER	MOTHER	DAUGHTER
BIFF			×
JIFF		×	×
SPIFF			

The remainder of the solution is revealed by the chart itself. It can now be seen that the only cell remaining opposite Jiff is under father. So, Jiff is the father. Show this by putting a ● in that cell:

	FATHER	MOTHER	DAUGHTER
BIFF	×		×
JIFF	●	×	×
SPIFF	×		

The chart subsequently reveals that Biff is the mother, since the only cell left opposite Biff is under the mother heading. So, put a ● in the appropriate cell, eliminating all other possibilities for mother with ×'s:

	Father	Mother	Daughter
Biff	×	●	×
Jiff	●	×	×
Spiff	×	×	

Finally, there is only one cell left opposite Spiff—under daughter:

	Father	Mother	Daughter
Biff	×	●	×
Jiff	●	×	×
Spiff	×	×	●

In summary, Biff is the mother, Jiff the father, and Spiff the daughter.

2 **Answer:** *Mr. Chow is the carpenter, Mr. Dow the painter, and Mr. Frow the plasterer.*

Solution: As in the solution of the previous puzzle, a two-axis chart will help you correlate the two sets of facts given in the puzzle—surnames (Chow, Dow, Frow) and occupations (carpenter, painter, plasterer):

	Carpenter	Painter	Plasterer
Chow			
Dow			
Frow			

You are told that: (1) the painter and the carpenter go out together on weekends; (2) the carpenter and the plasterer have known each other since childhood; and (3) Mr. Frow has never met Mr. Dow. From these statements, you can deduce, first and foremost, that the painter and the carpenter know each other, as do the carpenter and the plasterer:

Painter and Carpenter	Carpenter and Plasterer
↓	↓
Go out together on weekends	*Have known each other since childhood*

Obviously, the carpenter is the one who knows both the other two men—the painter and the plasterer. So, neither Mr. Frow nor Mr. Dow can be the carpenter, because these two men have never met each other. Consequently, you can safely eliminate carpenter as a possibility opposite both Dow and Frow:

	Carpenter	Painter	Plasterer
Chow			
Dow	×		
Frow	×		

That leaves Chow as the carpenter:

	Carpenter	Painter	Plasterer
Chow	●	×	×
Dow	×		
Frow	×		

You know that Mr. Dow and Mr. Frow are the plasterer and the painter, but not necessarily in that order. However, since you are also told that the plasterer makes more money than the painter, and that Mr. Frow makes more money than Mr. Dow, you can safely deduce that Mr. Frow is the plasterer and Mr. Dow the painter:

	Carpenter	Painter	Plasterer
Chow	●	×	×
Dow	×	●	×
Frow	×	×	●

In sum, Mr. Chow is the carpenter, Mr. Dow the painter, and Mr. Frow the plasterer.

3 **Answer:** *The boy dressed in red dated the girl dressed in blue; the boy dressed in green dated the girl dressed in red; and the boy dressed in blue dated the girl dressed in green.*

Solution: As before, set up a two-axis chart that will help you correlate two sets of facts—the colors worn by the boys (boy red, boy green, boy blue) and the colors worn by the girls (girl red, girl green, girl blue):

	Girl red	Girl green	Girl blue
Boy red			
Boy green			
Boy blue			

You are told by one of the boys that no one had a date with a partner dressed in the same color: that is, the boy dressed in red did not have a date with the girl dressed in red; the boy dressed in green did not have a date with the girl dressed in green; and the boy dressed in blue did not have a date with the girl dressed in blue:

	Girl red	Girl green	Girl blue
Boy red	×		
Boy green		×	
Boy blue			×

The boy who made this observation was dressed in red, and he was not dancing with the girl dressed in green, since he made the observation to her and her partner. So, you can safely eliminate girl green as a possibility for boy red:

	Girl red	Girl green	Girl blue
Boy red	×	×	
Boy green		×	
Boy blue			×

The chart now reveals simultaneously that the boy dressed in red dated the girl dressed in blue—since the only cell left opposite boy red is under girl blue—and that the boy dressed in blue dated the girl dressed in green—since the only cell left under girl green is opposite boy blue:

	Girl red	Girl green	Girl blue
Boy red	×	×	●
Boy green		×	×
Boy blue	×	●	×

So, the boy dressed in green dated the girl dressed in red—since the only cell left in the chart opposite boy green is under girl red:

	Girl red	Girl green	Girl blue
Boy red	×	×	●
Boy green	●	×	×
Boy blue	×	●	×

To summarize: the boy dressed in red dated the girl dressed in blue; the boy dressed in green dated the girl dressed in red; and the boy dressed in blue dated the girl dressed in green.

4 **Answer:** *Karen is the biologist, Lori the urban planner, Mary the physicist, Joan the geologist, Paula the meteorologist, and Peter the zoologist.*

Solution: This puzzle, like previous ones, also involves relating two sets of facts—names (Karen, Lori, Mary, Joan, Paula, Peter) and professions (geologist, meteorologist, physicist, urban planner, biologist, zoologist):

	Geologist	Meteorologist	Physicist	Urban planner	Biologist	Zoologist
Karen						
Lori						
Mary						
Joan						
Paula						
Peter						

Statement 1 tells you that Karen debated Lori and the meteorologist at the beginning of the conference. From this, it can be established that neither Karen nor Lori is the meteorologist:

	Geologist	Meteorologist	Physicist	Urban planner	Biologist	Zoologist
Karen		X				
Lori		X				
Mary						
Joan						
Paula						
Peter						

Statement 2 tells you that Peter is not the physicist, statement 3 that Mary is not the urban planner, and statement 4 that Joan is neither the meteorologist nor the biologist:

	Geologist	Meteorologist	Physicist	Urban planner	Biologist	Zoologist
Karen		X				
Lori		X				
Mary				X		
Joan		X			X	
Paula						
Peter			X			

Statement 5 identifies the six experts as the physicist, Karen, Joan, the zoologist, the female urban planner, and Paula. From this, you can deduce that: (1) Karen, Joan, and Paula are not the physicist, zoologist, or urban planner; and (2) Peter is not the urban planner, because the urban planner is female:

	Geologist	Meteorologist	Physicist	Urban planner	Biologist	Zoologist
Karen		×	×	×		×
Lori		×				
Mary				×		
Joan		×	×	×	×	×
Paula			×	×		×
Peter			×	×		

The chart by itself can now guide you the rest of the way. It reveals, first, that Lori is the urban planner, since the only cell left under urban planner is opposite Lori:

	Geologist	Meteorologist	Physicist	Urban planner	Biologist	Zoologist
Karen		×	×	×		×
Lori	×	×	×	●	×	×
Mary				×		
Joan		×	×	×	×	×
Paula			×	×		×
Peter			×	×		

Then, Mary is the physicist, since the only cell left under physicist is opposite Mary:

	Geologist	Meteorologist	Physicist	Urban planner	Biologist	Zoologist
Karen		×	×	×		×
Lori	×	×	×	●	×	×
Mary	×	×	●	×	×	×
Joan		×	×	×	×	×
Paula			×	×		×
Peter			×	×		

So, Peter is the zoologist, since the only cell left under zoologist is opposite Peter:

	Geologist	Meteorologist	Physicist	Urban planner	Biologist	Zoologist
Karen		×	×	×		×
Lori	×	×	×	●	×	×
Mary	×	×	●	×	×	×
Joan		×	×	×	×	×
Paula			×	×		×
Peter	×	×	×	×	×	●

Then Paula is the meteorologist, since the only cell left under meteorologist is opposite Paula:

	Geologist	Meteorologist	Physicist	Urban planner	Biologist	Zoologist
Karen		×	×	×		×
Lori	×	×	×	●	×	×
Mary	×	×	●	×	×	×
Joan		×	×	×	×	×
Paula	×	●	×	×	×	×
Peter	×	×	×	×	×	●

So, Joan is the geologist, since the only cell left opposite Joan is under geologist:

	Geologist	Meteorologist	Physicist	Urban planner	Biologist	Zoologist
Karen	×	×	×	×		×
Lori	×	×	×	●	×	×
Mary	×	×	●	×	×	×
Joan	●	×	×	×	×	×
Paula	×	●	×	×	×	×
Peter	×	×	×	×	×	●

Finally, Karen is the biologist, since the only cell left in the chart is under biologist and opposite Karen:

	Geologist	Meteorologist	Physicist	Urban planner	Biologist	Zoologist
Karen	×	×	×	×	●	×
Lori	×	×	×	●	×	×
Mary	×	×	●	×	×	×
Joan	●	×	×	×	×	×
Paula	×	●	×	×	×	×
Peter	×	×	×	×	×	●

In conclusion, Karen is the biologist, Lori the urban planner, Mary the physicist, Joan the geologist, Paula the meteorologist, and Peter the zoologist.

5 **Answer:** *Tam Grant is 6 years old, Larry Grant is 4 years old, Lori Winn is 3 years old, Rita Brown is 5 years old, and Gary Guild is 7 years old.*

Solution: This rather challenging puzzle requires you to relate three sets of facts—the children's names (Tam, Larry, Lori, Rita, Gary), surnames (Grant, Winn, Guild, Brown), and ages (3, 4, 5, 6, 7). Statement 1 informs you that Mrs.

Grant has more than one child. Since there are four surnames given in the puzzle—Grant, Winn, Guild, Brown—this means that she has exactly two children. Show this by using the surname Grant twice in the chart, which, like the chart used in illustrative example 2 above, will have to be set up with one of the sets of facts—say, the children's ages—repeated to the right and under the last names category:

		Last names					Ages				
		Grant	Grant	Winn	Guild	Brown	3	4	5	6	7
N	Tam										
A	Larry										
M	Lori										
E	Rita										
S	Gary										
A	3										
G	4										
E	5										
S	6										
	7										

Statement 1 tells you that every Saturday Mrs. Grant goes to work, leaving her two children with Mrs. Winn to take to the judo class with her, and that Mrs. Winn's daughter is younger than either one of Mrs. Grant's two children. From this, it can be established that: (1) neither Larry nor Gary is the Winn child, because Mrs. Winn has a daughter; (2) Mrs. Winn's daughter is not 6 or 7 years old, because she is younger than *both* of Mrs. Grant's children.

If you do not see this, then consider the situation in concrete terms. If Mrs. Winn's daughter were 7, then she would be older, not younger, than both of the Grant children. If she were 6, then one of the Grant children could be 7, but the other would necessarily have to be younger than 6. So, Mrs. Winn's daughter would not be younger than *both* of Mrs. Grant's children.

Register these deductions in the usual way:

		Last names					Ages				
		Grant	Grant	Winn	Guild	Brown	3	4	5	6	7
N	Tam										
A	Larry			×							
M	Lori										
E	Rita										
S	Gary			×							
A	3										
G	4										
E	5										
S	6			×							
	7			×							

Statement 2 tells you that Tam is older than Larry but younger than the Guild child. From this, it can be seen that: (1) neither Tam nor Larry is the Guild child; (2) Tam is neither 3 years old (for she is older than Larry) nor 7 years old (for she is younger than the Guild child); (3) Larry is not 7 (for Tam is older than he is); and (4) the Guild child is not 3 (for Tam is younger than that child).

Once again, show these deductions in the usual way:

		Last Names					Ages				
		Grant	Grant	Winn	Guild	Brown	3	4	5	6	7
Names	Tam				×		×				×
	Larry			×	×						×
	Lori										
	Rita										
	Gary			×							
Ages	3				×						
	4										
	5										
	6			×							
	7			×							

Statement 3 tells you that the Brown girl is 2 years older than Lori. From this, you can deduce that: (1) Lori is not the Brown child; (2) neither Larry nor Gary is the Brown child, since the Brown child is a girl; (3) Lori's age cannot be 6 or 7, since the Brown child is 2 years older than she is; and (4) the Brown child cannot be 3 or 4, since Lori is 2 years younger than the Brown child.

Once again, transfer these deductions to your chart in the usual way:

		Last Names					Ages				
		Grant	Grant	Winn	Guild	Brown	3	4	5	6	7
Names	Tam				×		×				×
	Larry			×	×	×					×
	Lori					×				×	×
	Rita										
	Gary			×		×					
Ages	3				×	×					
	4					×					
	5										
	6			×							
	7			×							

Statement 4 tells you that Rita's mother is sometimes home on Saturday, occasionally taking Gary to the class, while his mother goes shopping. From this, you can establish that: (1) Rita is not one of the Grant children, because her mother is home sometimes on Saturday, whereas from statement 1 you know that Mrs. Grant goes to work every Saturday; and (2) Gary is not one of the Grant children for the same reason—his mother goes shopping on Saturday, whereas Mrs. Grant works on that day.

Once again, show these deductions in the usual way:

		Last names					Ages				
		Grant	Grant	Winn	Guild	Brown	3	4	5	6	7
Names	Tam				×		×				×
	Larry			×	×	×					×
	Lori					×				×	×
	Rita	×	×								
	Gary	×	×	×		×					
Ages	3				×	×					
	4					×					
	5										
	6			×							
	7			×							

The chart now reveals that Gary is the Guild child, since the only cell left opposite Gary is under Guild:

		Last names					Ages				
		Grant	Grant	Winn	Guild	Brown	3	4	5	6	7
Names	Tam				×		×				×
	Larry			×	×	×					×
	Lori				×	×				×	×
	Rita	×	×		×						
	Gary	×	×	×	●	×					
Ages	3				×	×					
	4					×					
	5										
	6			×							
	7			×							

Then Larry is one of the Grant children—it doesn't matter which one in the chart—since the only two cells left opposite Larry are both under Grant:

		Last names					Ages				
		Grant	Grant	Winn	Guild	Brown	3	4	5	6	7
Names	Tam		×		×		×				×
	Larry	×	●	×	×	×					×
	Lori		×		×	×				×	×
	Rita	×	×		×						
	Gary	×	×	×	●	×					
Ages	3				×	×					
	4					×					
	5										
	6			×							
	7			×							

You can see in the lower age set that the Guild child is not 3; so, you can conclude that Gary, who is the Guild child, is not 3, showing this deduction with an × in the age set to the right opposite Gary. Statement 1 tells you that the Winn girl is younger than either Grant child. So, she is younger than Larry, who is one of the Grant children. This means that Larry cannot be 3 years old. Consequently, he is either 4 or 5—given that he cannot be 6 because you know that Tam is older than he is and that she is not 7 years old. Show these deductions with ×'s in age set to the right opposite Larry under both 3 and 6. Also, the fact that Larry is 4 or 5 entails that Tam can be only 5 or 6 (for she is older than he is). So, you can eliminate 4 years old as a possibility for Tam in the age set to the right:

		Last Names					Ages				
		Grant	Grant	Winn	Guild	Brown	3	4	5	6	7
Names	Tam		×		×		×	×			×
	Larry	×	●	×	×	×	×			×	×
	Lori		×		×	×				×	×
	Rita	×	×		×						
	Gary	×	×	×	●	×	×				
Ages	3				×	×					
	4					×					
	5										
	6			×							
	7			×							

Having established that Larry is one of the Grant children and that Gary is the Guild child, statement 2 can now be rephrased as: *Tam is older than Larry Grant but younger than Gary Guild.* You know that Tam is 5 or 6. So, Gary cannot be 4 because Tam is younger than he is. Show this by eliminating 4 as a possibility opposite Gary in the age set to the right and under Guild in the lower age set. Also, since Larry cannot be 3, 6, or 7, you can eliminate these possibilities under Grant in the lower age set:

		Last Names					Ages				
		Grant	Grant	Winn	Guild	Brown	3	4	5	6	7
Names	Tam		×		×		×	×			×
	Larry	×	●	×	×	×	×			×	×
	Lori		×		×	×				×	×
	Rita	×	×		×						
	Gary	×	×	×	●	×	×	×			
Ages	3		×		×	×					
	4				×	×					
	5										
	6		×	×							
	7		×	×							

You can also eliminate 3 years old in the lower set as a possibility under the first Grant, because from statement 1 you know that neither Grant child is the youngest child, given that Mrs. Winn's daughter is younger than both of the Grant children:

		LAST NAMES					AGES				
		GRANT	GRANT	WINN	GUILD	BROWN	3	4	5	6	7
N	TAM		×		×		×	×			×
A	LARRY	×	●	×	×	×	×			×	×
M	LORI		×		×	×				×	×
E	RITA	×	×		×						
S	GARY	×	×	×	●	×	×	×			
A	3	×	×		×	×					
G	4				×	×					
E	5										
S	6		×	×							
	7		×	×							

The chart now shows that the Winn child is 3 years old, since the only cell left opposite 3 in the lower age set is under Winn:

		LAST NAMES					AGES				
		GRANT	GRANT	WINN	GUILD	BROWN	3	4	5	6	7
N	TAM		×		×		×	×			×
A	LARRY	×	●	×	×	×	×			×	×
M	LORI		×		×	×				×	×
E	RITA	×	×		×						
S	GARY	×	×	×	●	×	×	×			
A	3	×	×	●	×	×					
G	4			×	×	×					
E	5			×							
S	6		×	×							
	7		×	×							

This implies that Tam cannot be the Winn child, because you have established already that she is not 3 years old:

		LAST NAMES					AGES				
		GRANT	GRANT	WINN	GUILD	BROWN	3	4	5	6	7
N	TAM		×	×	×		×	×			×
A	LARRY	×	●	×	×	×	×			×	×
M	LORI		×		×	×				×	×
E	RITA	×	×		×						
S	GARY	×	×	×	●	×	×	×			
A	3	×	×	●	×	×					
G	4			×	×	×					
E	5			×							
S	6		×	×							
	7		×	×							

You know, on the one hand, that Gary is not a Grant child—he is the Guild child, as you found out—and that Rita's mother occasionally takes Gary to class on Saturday (statement 4). On the other hand, you know that it is Mrs. Winn who takes the Grant children to class on Saturday and that Rita's mother does not (statement 4). So, you can safely say that Rita's mother is not Mrs. Winn, eliminating Winn as a possibility opposite Rita in the chart:

		LAST NAMES					AGES				
		GRANT	GRANT	WINN	GUILD	BROWN	3	4	5	6	7
NAMES	TAM		×	×	×		×	×			×
	LARRY	×	●	×	×	×	×			×	×
	LORI		×		×	×				×	×
	RITA	×	×	×	×						
	GARY	×	×	×	●	×	×	×			
AGES	3	×	×	●	×	×					
	4			×	×	×					
	5			×							
	6		×	×							
	7		×	×							

The chart now reveals simultaneously that Rita is the Brown child, since the only cell left opposite Rita is under Brown, and that Lori is the Winn child, since the only cell left under Winn is opposite Lori:

		LAST NAMES					AGES				
		GRANT	GRANT	WINN	GUILD	BROWN	3	4	5	6	7
NAMES	TAM		×	×	×	×	×	×			×
	LARRY	×	●	×	×	×	×			×	×
	LORI	×	×	●	×	×				×	×
	RITA	×	×	×	×	●					
	GARY	×	×	×	●	×	×	×			
AGES	3	×	×	●	×	×					
	4			×	×	×					
	5			×							
	6		×	×							
	7		×	×							

It can now be seen that Tam is the other Grant child, since the only cell left opposite her name is under Grant:

		Last names					Ages				
		Grant	Grant	Winn	Guild	Brown	3	4	5	6	7
Names	Tam	●	×	×	×	×	×	×			×
	Larry	×	●	×	×	×	×			×	×
	Lori	×	×	●	×	×				×	×
	Rita	×	×	×	×	●					
	Gary	×	×	×	●	×	×	×			
Ages	3	×	×	●	×	×					
	4			×	×	×					
	5			×							
	6		×	×							
	7		×	×							

You can now complete the rest of the chart easily. Lori is 3 years old, because she is the Winn child, and you have discovered that the Winn child is the 3-year-old. Also, you can eliminate the ages 3, 4, and 7 in the lower set under Grant, because you know that Tam, who is that Grant child, is either 5 or 6:

		Last names					Ages				
		Grant	Grant	Winn	Guild	Brown	3	4	5	6	7
Names	Tam	●	×	×	×	×	×	×			×
	Larry	×	●	×	×	×	×			×	×
	Lori	×	×	●	×	×	●	×	×	×	×
	Rita	×	×	×	×	●	×				
	Gary	×	×	×	●	×	×	×			
Ages	3	×	×	●	×	×					
	4	×		×	×	×					
	5			×							
	6		×	×							
	7	×	×	×							

The lower age set then reveals that Larry Grant is the 4-year-old, since the only cell left under the second Grant is opposite 4. Show this deduction as well opposite Larry under 4 in the age set to the right:

		Last names					Ages				
		Grant	Grant	Winn	Guild	Brown	3	4	5	6	7
Names	Tam	●	×	×	×	×	×	×			×
	Larry	×	●	×	×	×	×	●	×	×	×
	Lori	×	×	●	×	×	●	×	×	×	×
	Rita	×	×	×	×	●	×	×			
	Gary	×	×	×	●	×	×	×			
Ages	3	×	×	●	×	×					
	4	×	●	×	×	×					
	5		×	×							
	6		×	×							
	7	×	×	×							

Note that Lori Winn is 3 years old, and that the Brown child—Rita Brown—is 2 years older (statement 3). So, she is 5 years old. Show this in both the lower age set and the one to the right:

		LAST NAMES					AGES				
		GRANT	GRANT	WINN	GUILD	BROWN	3	4	5	6	7
NAMES	TAM	●	×	×	×	×	×	×	×		×
	LARRY	×	●	×	×	×	×	●	×	×	×
	LORI	×	×	●	×	×	●	×	×	×	×
	RITA	×	×	×	×	●	×	×	●	×	×
	GARY	×	×	×	●	×	×	×	×		
AGES	3	×	×	●	×	×					
	4	×	●	×	×	×					
	5	×	×	×	×	●					
	6		×	×		×					
	7	×	×	×		×					

The chart now reveals that Tam Grant is 6 years old, since the only cell left under Grant in the lower age set is opposite 6, and the only cell left in the age set to the right opposite Tam is under age 6:

		LAST NAMES					AGES				
		GRANT	GRANT	WINN	GUILD	BROWN	3	4	5	6	7
NAMES	TAM	●	×	×	×	×	×	×	×	●	×
	LARRY	×	●	×	×	×	×	●	×	×	×
	LORI	×	×	●	×	×	●	×	×	×	×
	RITA	×	×	×	×	●	×	×	●	×	×
	GARY	×	×	×	●	×	×	×	×	×	
AGES	3	×	×	●	×	×					
	4	×	●	×	×	×					
	5	×	×	×	×	●					
	6	●	×	×	×	×					
	7	×	×	×		×					

This leaves Gary Guild as the 7-year-old:

		LAST NAMES					AGES				
		GRANT	GRANT	WINN	GUILD	BROWN	3	4	5	6	7
NAMES	TAM	●	×	×	×	×	×	×	×	●	×
	LARRY	×	●	×	×	×	×	●	×	×	×
	LORI	×	×	●	×	×	●	×	×	×	×
	RITA	×	×	×	×	●	×	×	●	×	×
	GARY	×	×	×	●	×	×	×	×	×	●
AGES	3	×	×	●	×	×					
	4	×	●	×	×	×					
	5	×	×	×	×	●					
	6	●	×	×	×	×					
	7	×	×	×	●	×					

In sum, Tam Grant is 6 years old, Larry Grant is 4 years old, Lori Winn is 3 years old, Rita Brown is 5 years old, and Gary Guild is 7 years old.

2 Puzzles in Truth Logic

"What is truth?" Throughout history, this question has intrigued great religious leaders, philosophers, artists, writers, and scientists. But it is also a more modest question that puzzlists love to pose. In fact, there exists a popular variety of puzzles that plays artfully on what logicians call "truth conditions."

Puzzles that fall within this genre typically contain statements, a certain number of which are known to be true and others false. Your task is to get to the "truth of the matter," *logically speaking,* of course.

How To...

There are two types of truth logic puzzles that will concern us in this chapter. The first is illustrated in example 1 below; it involves identifying *who* did *what* (steal something, murder someone, etc.) on the basis of certain statements made by various people, some of which are true and others false. The second type, which is illustrated in example 2 below, involves determining the group, tribe, and so on, to which an individual, or small group of people, belong on the basis of certain statements that are made.

Puzzle Properties

Like the puzzles in the previous chapter, the two types of puzzles to be dealt with here involve no play on words, no guessing, no technical know-how—only the ability to think clearly and work methodically. A puzzle belonging to the first type will invariably contain statements made by individuals (suspects in a murder case, suspects in a robbery case, etc.), some of which are true (T) and others false (F). For example, you might be told that certain bank robbery suspects made three statements each: *I did it, He didn't do it, I don't know the other suspect*—of which one was false (=1F) and two true (=2T). Your task is to establish who the thief is, by working out a "truth-value" arrangement of 1F and 2T's per set of statements that does not lead to a logical inconsistency.

Puzzles of the second type require that you identify to what tribe or group a certain individual (or individuals) belong(s) on the basis of statements he or she makes or that some other person makes. For example, you might be told that two members of a tribe belong to different families, one of which is known always to tell the truth and the other always to lie. On the basis of statements made by each one, your task will be to establish who belongs to which family.

Example 1 The following is an example of a truth logic puzzle of the first type.

Billy Bones was found murdered one night in an alley behind the night club he usually frequented. The police caught three suspects the morning after. That afternoon, the three men were interrogated by a police investigator. They made the following statements:

Ben:
1. I didn't kill Billy.
2. Jim is not my friend.
3. I knew Billy.

Jim:
1. I didn't kill Billy.
2. Ben and Tim are friends of mine.
3. Ben didn't kill Billy.

Tim:
1. I didn't kill Billy.
2. Ben lied when he said that Jim was not his friend.
3. I don't know who killed Billy.

Only one of the three is guilty, and only one of each man's statements is false. Who killed Billy Bones?

Like the puzzles in the previous chapter, it would be a mind-boggling task to keep track of all the possible true–false arrangements that this puzzle implies without some visual organizing device to help you keep track of them. So, the first thing to do is to set up a *truth chart* in which you can display the statements of all three men and register their truth values as you go along:

	Statement	Truth value
Ben	1. I didn't kill Billy. 2. Jim is not my friend. 3. I knew Billy.	1. 2. 3.
Jim	1. I didn't kill Billy. 2. Ben and Tim are friends of mine. 3. Ben didn't kill Billy.	1. 2. 3.
Tim	1. I didn't kill Billy. 2. Ben lied when he said that Jim was not his friend. 3. I don't know who killed Billy.	1. 2. 3.

Putting T's (true statements) and F's (false statements) in this simple chart will allow you to keep track of the facts as you discover them:

- If you conclude, for instance, that one of the three statements a suspect makes is definitely his false one, then place an F opposite that statement.

- If, however, you deduce that one of his three statements cannot be his false one, then place a T opposite that statement.
- The solution is complete when you have placed exactly one F and two T's opposite each man's three statements successfully. Do you see why? The puzzle tells you exactly how many F's and T's are to be assigned to each set of three statements by informing you that *only one of each man's statements is false* (=1F). This means, of course, that his other two are true (=2T). So, the solution is complete when you have placed 1F and 2T's successfully into each set of statements: that is, the solution will emerge when you have entered 1F and 2T's in a consistent, noncontradictory way opposite the three statements made by each suspect.

The first thing to note is that the first statement uttered by each suspect is the same—*I didn't kill Billy.* Since we are told that one of the three is the murderer, then one of these statements is necessarily false (F). If they were all true, then there would be no murderer! Assume that Ben's first statement is the false one. Then, the first statements of the other two suspects are necessarily true (remember that there can be only one murderer). So, go ahead and register an F opposite Ben's first statement and T's opposite the corresponding statements of the other two suspects:

	STATEMENT	TRUTH VALUE
BEN	1. I DIDN'T KILL BILLY. 2. JIM IS NOT MY FRIEND. 3. I KNEW BILLY.	1. F 2. 3.
JIM	1. I DIDN'T KILL BILLY. 2. BEN AND TIM ARE FRIENDS OF MINE. 3. BEN DIDN'T KILL BILLY.	1. T 2. 3.
TIM	1. I DIDN'T KILL BILLY. 2. BEN LIED WHEN HE SAID THAT JIM WAS NOT HIS FRIEND. 3. I DON'T KNOW WHO KILLED BILLY.	1. T 2. 3.

Now, follow the logical implications of your assumption to see where they lead. Start your logical journey by adding T's opposite Ben's second and third statements, since you know that only one of his three statements is false, while his other two are true:

	STATEMENT	TRUTH VALUE
BEN	1. I DIDN'T KILL BILLY. 2. JIM IS NOT MY FRIEND. 3. I KNEW BILLY.	1. F 2. T 3. T
JIM	1. I DIDN'T KILL BILLY. 2. BEN AND TIM ARE FRIENDS OF MINE. 3. BEN DIDN'T KILL BILLY.	1. T 2. 3.
TIM	1. I DIDN'T KILL BILLY. 2. BEN LIED WHEN HE SAID THAT JIM WAS NOT HIS FRIEND. 3. I DON'T KNOW WHO KILLED BILLY.	1. T 2. 3.

Note that Ben's second statement—*Jim is not my friend*—now has a T-value. Consider Jim's second statement—*Ben and Tim are friends of mine.* Is it true or false? Well, it clearly contradicts Ben's second statement, since Ben says that Jim is not his friend. So, Jim's second statement is apparently false. Similarly, consider Tim's second statement—*Ben lied when he said that Jim was not his friend.* It too is false. Do you see why? Consider Ben's second statement again—*Jim is not my friend.* The chart indicates that this is a true statement. So, according to the chart, Ben did not lie when he said that Jim was not his friend, as Tim asserts. The liar is, therefore, Tim. Register these two new F's in the chart—one opposite Jim's second statement and one opposite Tim's second statement:

	Statement	Truth value
Ben	1. I didn't kill Billy. 2. Jim is not my friend. 3. I knew Billy.	1. F 2. T 3. T
Jim	1. I didn't kill Billy. 2. Ben and Tim are friends of mine. 3. Ben didn't kill Billy.	1. T 2. F 3.
Tim	1. I didn't kill Billy. 2. Ben lied when he said that Jim was not his friend. 3. I don't know who killed Billy.	1. T 2. F 3.

It is now an easy task to complete the chart. You know that each suspect made one false (1F) and two true (2T's) statements. Look at Jim's and Tim's three statements, and you will see that each one has made 1F and 1T statement according to the chart. So, go ahead and add the remaining T's to Jim's and Tim's set of statements—namely, opposite both of their third statements:

	Statement	Truth value
Ben	1. I didn't kill Billy. 2. Jim is not my friend. 3. I knew Billy.	1. F 2. T 3. T
Jim	1. I didn't kill Billy. 2. Ben and Tim are friends of mine. 3. Ben didn't kill Billy.	1. T 2. F 3. T
Tim	1. I didn't kill Billy. 2. Ben lied when he said that Jim was not his friend. 3. I don't know who killed Billy.	1. T 2. F 3. T

So, who is the murderer? Consider Ben's first statement—*I didn't kill Billy*—noting that it has an F-value. That is the assumption with which you started your logical journey through the arrangement of F's and T's. Now that you've completed the chart, what can you infer from Ben's first statement? If it is false that he didn't kill Billy, then he did indeed kill Billy. Ben is therefore our murderer, or is he? Consider Jim's third statement—*Ben didn't kill Billy.* According to the chart, it is a true statement. So, according to Jim's assertion, *which is true according to the chart,* Ben is not the murderer! So, is he or isn't he the murderer? Obviously, you cannot tell from this particular arrangement of F's and T's. It is an arrangement that turns out to be logically *inconsistent.* What caused the inconsistency? It can only be the assumption with which you started out on your logical journey—namely, that Ben's first statement is his false one. Like a virus in a computer, that assumption must be rooted out! Ben's first statement, therefore, is definitely not his false one. As a corollary, it is necessarily one of his two true ones. So, go ahead and register this sure finding in your chart, erasing all the other F's and T's from it:

	STATEMENT	TRUTH VALUE
BEN	1. I DIDN'T KILL BILLY. 2. JIM IS NOT MY FRIEND. 3. I KNEW BILLY.	1. T 2. 3.
JIM	1. I DIDN'T KILL BILLY. 2. BEN AND TIM ARE FRIENDS OF MINE. 3. BEN DIDN'T KILL BILLY.	1. 2. 3.
TIM	1. I DIDN'T KILL BILLY. 2. BEN LIED WHEN HE SAID THAT JIM WAS NOT HIS FRIEND. 3. I DON'T KNOW WHO KILLED BILLY.	1. 2. 3.

Since Ben's first statement—*I didn't kill Billy*—has now been established as being true, you can see that Jim's third statement is also true—*Ben didn't kill Billy*—because it simply confirms what Ben said in his first statement. So, you can safely put a T-value in the chart opposite Jim's third statement:

	STATEMENT	TRUTH VALUE
BEN	1. I DIDN'T KILL BILLY. 2. JIM IS NOT MY FRIEND. 3. I KNEW BILLY.	1. T 2. 3.
JIM	1. I DIDN'T KILL BILLY. 2. BEN AND TIM ARE FRIENDS OF MINE. 3. BEN DIDN'T KILL BILLY.	1. 2. 3. T
TIM	1. I DIDN'T KILL BILLY. 2. BEN LIED WHEN HE SAID THAT JIM WAS NOT HIS FRIEND. 3. I DON'T KNOW WHO KILLED BILLY.	1. 2. 3.

Assume that Tim's first statement is his false one, putting an F opposite that statement and adding his 2T's opposite his second and third statements. Register these hypothetical F and T's in a different-colored pen or pencil to distinguish them from the two sure T-values you have established so far—namely, the T opposite Ben's first statement and the T opposite Jim's third statement (here we have used italic type):

	STATEMENT	TRUTH VALUE
BEN	1. I DIDN'T KILL BILLY. 2. JIM IS NOT MY FRIEND. 3. I KNEW BILLY.	1. T 2. 3.
JIM	1. I DIDN'T KILL BILLY. 2. BEN AND TIM ARE FRIENDS OF MINE. 3. BEN DIDN'T KILL BILLY.	1. 2. 3. T
TIM	1. I DIDN'T KILL BILLY. 2. BEN LIED WHEN HE SAID THAT JIM WAS NOT HIS FRIEND. 3. I DON'T KNOW WHO KILLED BILLY.	1. *F* 2. *T* 3. *T*

Where does this new assumption lead? First and foremost, it implies that Tim is the murderer, because according to the chart, Tim's first statement—*I didn't kill Billy*—is false. So, he did indeed kill Billy. Now, consider Jim's first statement—*I didn't kill Billy.* Since there is only one murderer, Tim, it is obvious that Jim did not kill Billy. His first statement simply confirms this. So, it is true. Go ahead and register a T-value in a different color opposite Jim's first statement:

	STATEMENT	TRUTH VALUE
BEN	1. I DIDN'T KILL BILLY.	1. T
	2. JIM IS NOT MY FRIEND.	2.
	3. I KNEW BILLY.	3.
JIM	1. I DIDN'T KILL BILLY.	1. *T*
	2. BEN AND TIM ARE FRIENDS OF MINE.	2.
	3. BEN DIDN'T KILL BILLY.	3. T
TIM	1. I DIDN'T KILL BILLY.	1. *F*
	2. BEN LIED WHEN HE SAID THAT JIM WAS NOT HIS FRIEND.	2. *T*
	3. I DON'T KNOW WHO KILLED BILLY.	3. *T*

Recall that for each set of three statements, there must be 1F and 2T's. Look at Jim's set of statements. Note that there are 2T's in it—one opposite his first statement and one opposite his third statement. This means that Jim's second statement must be assigned an F-value:

	STATEMENT	TRUTH VALUE
BEN	1. I DIDN'T KILL BILLY.	1. T
	2. JIM IS NOT MY FRIEND.	2.
	3. I KNEW BILLY.	3.
JIM	1. I DIDN'T KILL BILLY.	1. *T*
	2. BEN AND TIM ARE FRIENDS OF MINE.	2. *F*
	3. BEN DIDN'T KILL BILLY.	3. T
TIM	1. I DIDN'T KILL BILLY.	1. *F*
	2. BEN LIED WHEN HE SAID THAT JIM WAS NOT HIS FRIEND.	2. *T*
	3. I DON'T KNOW WHO KILLED BILLY.	3. *T*

Now, focus your attention on Ben's three statements. You know that his first one is true. You established that fact a while ago. Now, consider his second statement—*Jim is not my friend.* Is it true or false? Tim's second statement—*Ben lied when he said that Jim was not his friend*—can help you determine that. It has a T-value. That means that when Ben said *Jim is not my friend,* he in fact lied. So, Ben's second statement must be false. Indicate this in a different color as well:

	STATEMENT	TRUTH VALUE
BEN	1. I DIDN'T KILL BILLY.	1. T
	2. JIM IS NOT MY FRIEND.	2. *F*
	3. I KNEW BILLY.	3.
JIM	1. I DIDN'T KILL BILLY.	1. *T*
	2. BEN AND TIM ARE FRIENDS OF MINE.	2. *F*
	3. BEN DIDN'T KILL BILLY.	3. T
TIM	1. I DIDN'T KILL BILLY.	1. *F*
	2. BEN LIED WHEN HE SAID THAT JIM WAS NOT HIS FRIEND.	2. *T*
	3. I DON'T KNOW WHO KILLED BILLY.	3. *T*

Remember that each suspect made only one false statement. The chart shows that Ben made one true (his first) and one false (his second) statement. So, his third statement is necessarily true:

	STATEMENT	TRUTH VALUE
BEN	1. I DIDN'T KILL BILLY. 2. JIM IS NOT MY FRIEND. 3. I KNEW BILLY.	1. T 2. *F* 3. *T*
JIM	1. I DIDN'T KILL BILLY. 2. BEN AND TIM ARE FRIENDS OF MINE. 3. BEN DIDN'T KILL BILLY.	1. *T* 2. *F* 3. T
TIM	1. I DIDN'T KILL BILLY. 2. BEN LIED WHEN HE SAID THAT JIM WAS NOT HIS FRIEND. 3. I DON'T KNOW WHO KILLED BILLY.	1. *F* 2. *T* 3. *T*

The chart is now complete again. Is this new arrangement of F's and T's logically consistent? Let's find out. Consider Jim's second statement—*Ben and Tim are friends of mine.* What does it imply? Since it is false, it implies that Ben is not one of Jim's friends. So, Ben's second statement—*Jim is not my friend*—is in fact true. But, according to the chart, it has an F-value! So, is his statement true or false? Obviously, you cannot determine its truth-value on the basis of this new arrangement of F's and T's. Once again, the arrangement turns out to be logically inconsistent. What caused the inconsistency this time? It can only be the assumption with which you started out on your second logical journey—namely, that Tim's first statement is his false one. As it turns out, it is definitely not his false one. It is therefore necessarily one of his two true ones. So, first erase all the F's and T's in a different color from the chart, leaving in it only the two previous T-values that were established during your first logical journey. Then, go ahead and register a new T-value opposite Tim's first statement:

	STATEMENT	TRUTH VALUE
BEN	1. I DIDN'T KILL BILLY. 2. JIM IS NOT MY FRIEND. 3. I KNEW BILLY.	1. T 2. 3.
JIM	1. I DIDN'T KILL BILLY. 2. BEN AND TIM ARE FRIENDS OF MINE. 3. BEN DIDN'T KILL BILLY.	1. 2. 3. T
TIM	1. I DIDN'T KILL BILLY. 2. BEN LIED WHEN HE SAID THAT JIM WAS NOT HIS FRIEND. 3. I DON'T KNOW WHO KILLED BILLY.	1. T 2. 3.

Incidentally, you have now identified the murderer. It is, of course, Jim, whose first statement—*I didn't kill Billy*—has been confirmed by the process of elimination as the false one of the three first statements made by the suspects. Nevertheless, it is always wise to complete the chart, just to make sure that no inconsistencies arise from this finding. So, start your third logical journey by putting an F-value opposite Jim's first statement:

	Statement	Truth value
Ben	1. I didn't kill Billy. 2. Jim is not my friend. 3. I knew Billy.	1. T 2. 3.
Jim	1. I didn't kill Billy. 2. Ben and Tim are friends of mine. 3. Ben didn't kill Billy.	1. F 2. 3. T
Tim	1. I didn't kill Billy. 2. Ben lied when he said that Jim was not his friend. 3. I don't know who killed Billy.	1. T 2. 3.

Since his third statement has a T-value, then his second statement is necessarily his other true statement:

	Statement	Truth value
Ben	1. I didn't kill Billy. 2. Jim is not my friend. 3. I knew Billy.	1. T 2. 3.
Jim	1. I didn't kill Billy. 2. Ben and Tim are friends of mine. 3. Ben didn't kill Billy.	1. F 2. T 3. T
Tim	1. I didn't kill Billy. 2. Ben lied when he said that Jim was not his friend. 3. I don't know who killed Billy.	1. T 2. 3.

Ben's second statement can now be seen to be false. Why? Because Jim's second statement—*Ben and Tim are friends of mine*—makes it false. Jim's statement has a T-value, implying that Ben and Tim are indeed Jim's friends, or, to put it another way, that all three—Ben, Jim, and Tim—are friends. So, when Ben says *Jim is not my friend,* he is clearly lying. Put an F-value opposite Ben's second statement:

	Statement	Truth value
Ben	1. I didn't kill Billy. 2. Jim is not my friend. 3. I knew Billy.	1. T 2. F 3.
Jim	1. I didn't kill Billy. 2. Ben and Tim are friends of mine. 3. Ben didn't kill Billy.	1. F 2. T 3. T
Tim	1. I didn't kill Billy. 2. Ben lied when he said that Jim was not his friend. 3. I don't know who killed Billy.	1. T 2. 3.

Complete Ben's set of statements by adding his missing T-value opposite his third statement. Note that this does not lead to any contradiction. It simply asserts that Ben knew the victim, Billy:

	Statement	Truth value
Ben	1. I didn't kill Billy.	1. T
	2. Jim is not my friend.	2. F
	3. I knew Billy.	3. T
Jim	1. I didn't kill Billy.	1. F
	2. Ben and Tim are friends of mine.	2. T
	3. Ben didn't kill Billy.	3. T
Tim	1. I didn't kill Billy.	1. T
	2. Ben lied when he said that Jim was not his friend.	2.
	3. I don't know who killed Billy.	3.

Consider again Ben's second statement—*Jim is not my friend*—which you have established as being false. It allows you to conclude that Tim's second statement—*Ben lied when he said that Jim was not his friend*—is true. Why? Because the chart shows that Ben lied when he said that Jim was not his friend (=Ben's second statement). This is exactly what Tim tells us with his own second statement. You can now safely put a T-value opposite Tim's second statement:

	Statement	Truth value
Ben	1. I didn't kill Billy.	1. T
	2. Jim is not my friend.	2. F
	3. I knew Billy.	3. T
Jim	1. I didn't kill Billy.	1. F
	2. Ben and Tim are friends of mine.	2. T
	3. Ben didn't kill Billy.	3. T
Tim	1. I didn't kill Billy.	1. T
	2. Ben lied when he said that Jim was not his friend.	2. T
	3. I don't know who killed Billy.	3.

This means that Tim's third statement is necessarily false:

	Statement	Truth value
Ben	1. I didn't kill Billy.	1. T
	2. Jim is not my friend.	2. F
	3. I knew Billy.	3. T
Jim	1. I didn't kill Billy.	1. F
	2. Ben and Tim are friends of mine.	2. T
	3. Ben didn't kill Billy.	3. T
Tim	1. I didn't kill Billy.	1. T
	2. Ben lied when he said that Jim was not his friend.	2. T
	3. I don't know who killed Billy.	3. F

So, contrary to what he says—*I don't know who killed Billy*—Tim did indeed know who killed the victim, Billy. But this contradicts nothing we have discovered so far. In summary, since this third arrangement of F's and T's leads to no logical inconsistencies, it can be asserted that Jim is indeed the one who killed Billy.

Example 2 The second type of truth logic puzzle that will concern us in this chapter normally involves determining *which* individuals belong to *which* tribe, clan, group, and so on, on the basis of certain statements that are made. The following is a version of a classic nut in this genre.

> The people of an island culture belong to one of two tribes—the Bawi or the Mawi. Since they look and dress alike, and since they speak the same language, they are virtually indistinguishable. It is known, however, that the members of the Bawi tribe always tell the truth, whereas the members of the Mawi tribe always lie. The anthropologist who became interested in studying their fascinating social system, Dr. Mary Titherington, recently came across three male individuals.
>
> "To which tribe do you belong?" Dr. Titherington asked the first individual.
>
> "Doo-too looh-nooh," replied the individual in his native language.
>
> "What did he say?" asked the anthropologist of the second and third individuals, both of whom had learned to speak some English.
>
> "He said that he is a Bawi," said the second.
>
> "No, he said that he is a Mawi," said the third.
>
> Can you figure out to what tribes the second and third individuals belonged?

The key to finding a solution to this, and to all puzzles of this type, is to zero in on a specific statement in order to test its truth-value or, as in this case, to unravel its actual content (i.e., what it really says). The key to solving this particular puzzle is to translate *Doo-too looh-nooh* into English. How do you go about doing this? Assume that the first individual belonged to the Bawi tribe. You are told that the members of the Bawi tribe always tell the truth. So, his answer in English to the anthropologist's question *To which tribe do you belong?* would have been, of course, that he belonged to the Bawi tribe. If he had said that he belonged to the Mawis, he would have been lying! So, in this hypothetical scenario, *Doo-too looh-nooh* translates as *I belong to the Bawi tribe:*

➤ **Scenario 1: First Individual = A Bawi**

Dr. Titherington: *To which tribe do you belong?*
Bawi Individual: *Doo-too looh-nooh = I belong to the Bawi tribe.*
(which is true, since Bawis always tell the truth)

Now assume the opposite scenario: namely, that the first individual belonged to the Mawi tribe. You are told that the members of that tribe always lie. So, his answer to Titherington's question—*To which tribe do you belong?*—would have been a lie. So, what was his answer? He certainly would not have admitted to our anthropologist that he belonged to the Mawi tribe. Instead, according to his nature he would have lied, saying that he belonged to the other tribe, the Bawis. So, his answer in English would have been that he too belonged to the Bawi tribe. Once again, in this second hypothetical scenario *Doo-too looh-nooh* translates as *I belong to the Bawi tribe:*

➤ **Scenario 2: First Individual = A Mawi**

Dr. Titherington: *To which tribe do you belong?*
Mawi Individual: *Doo-too looh-nooh = I belong to the Bawi tribe.* (which is false, because a Mawi would never admit that he was in fact a Mawi)

As you can see, no matter to which tribe the first individual really belonged, the anthropologist would have gotten the same answer from him to her question *To which tribe do you belong?*—namely, *I belong to the Bawi tribe.* Now, consider the responses given by the other two individuals to Titherington's follow-up question—*What did he say?* Start with the second individual's response:

Dr. Titherington: *What did he say?*
Second Individual: *He said that he is a Bawi.*

As you have just discovered, the first individual did indeed say that he was a Bawi. So, this second individual told the truth. What does that imply? Well, of course, that he himself is a member of the Bawi tribe.

Finally, consider the response given by the third individual to Titherington's follow-up question:

Dr. Titherington: *What did he say?*
Third Individual: *He said that he is a Mawi.*

As you know, the first individual said that he was a Bawi. So, this third individual clearly lied. What does that imply? That he is, of course, a member of the mendacious Mawi tribe.

Summary

The main line of attack in solving truth puzzles of the type illustrated in example 1 above is to set up a truth chart that displays all the statements made by the various people and in which you can register the truth-values to be assigned to them as you go along. The idea is to put T's and F's in the chart according to some working assumption. If the assumption leads to an inconsistency, then you must discard it, starting again with any relevant information you have gained in the process duly recorded in the chart.

In summary, when solving such puzzles, do the following:

- Set up a truth chart so that you can keep track of the various arrangements of T-values and F-values that can be assumed or inferred.
- Assign T's and F's to each set of statements in the chart as required by the conditions set out by the puzzle.
- Start by assuming that a certain statement is either true (T) or false (F), following your assumption right through to its logical consequences. If it leads to an inconsistency, then you have discovered that the opposite truth-value is

the tenable one: that is, if you assumed that a statement was true, then it has turned out really to be false, and vice versa. If it does not lead to an inconsistency, then you have discovered its actual truth-value: that is, your assumption that a statement was true or false has, in fact, turned out to be a tenable one.

- When the arrangement of T's and F's leads to a complete system of consistent statements, you will have then solved the puzzle.

To solve puzzles of the type illustrated in example 2, start by considering either the truth-value or actual content of a specific statement. Examine the other statements in relation to the selected statement, testing their consistency or validity against its truth-value or content. This line of attack will generally yield the puzzle's solution.

Puzzles 6–10

Answers, along with step-by-step solutions, can be found at the end of the chapter.

6 Jim, Bud, and Sam were rounded up by the police yesterday, because one of them was suspected of having robbed the local bank. The three suspects made the following statements under intensive questioning:

Jim: *I'm innocent.*
Bud: *I'm innocent.*
Sam: *Bud is the guilty one.*

If only one of these statements turned out to be true, who robbed the bank?

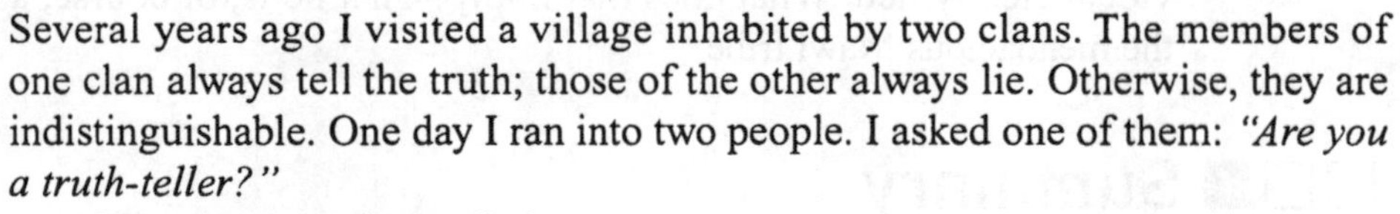

7 Several years ago I visited a village inhabited by two clans. The members of one clan always tell the truth; those of the other always lie. Otherwise, they are indistinguishable. One day I ran into two people. I asked one of them: *"Are you a truth-teller?"*

"Ooma tooma," she replied.

I asked her partner, a tall young man who spoke English, what she had said.

"She said no," was his reply.

Can you figure out to which clan the young man belonged?

8 The next day, in another part of the same village, where the members of one clan always tell the truth and those of the other always lie, I ran into two other people—a female and a male. However, they spoke a different dialect of the village language than the one spoken by my previous informants.

I asked the female, as I had asked the other one, *"Are you a truth-teller?"*

"Goo boo," she replied in her dialect.

I asked her partner, a short man who spoke English, what she had said.

"She said yes," was his reply, *"but she is a liar."*

Can you figure out to which clan(s) both my informants belonged?

9 Anita, Brunhilde, Claudia, Daniela, Ella, and Frida are good friends who love to argue literally over the time of day! Here they are again arguing mischievously over what day of the week it is:

Anita: *The day before yesterday was Friday or Saturday.*

Brunhilde: *No, you're wrong. Today is Saturday.*

Claudia: *No, today is not Saturday. Nor is it Sunday or Monday.*

Daniela: *The day after tomorrow is Tuesday.*

Ella: *Tomorrow cannot be Friday or Saturday.*

Frida: *Tomorrow is Friday.*

Only one of their statements is true. All the others are false. Can you determine which day of the week it is?

10 George, Harry, and Michael are being interviewed for the position of associate puzzle editor for a famous puzzle magazine. Each man made statements to the job interviewer that revealed the wit of an inveterate puzzlist. Here is a sample of what each one said:

George:
1. *I'm twenty-two.*
2. *Harry is two years older than I am.*
3. *Michael is one year younger than I am.*

Harry:
1. *I am not the youngest.*
2. *There is a three-year difference between my age and Michael's age.*
3. *Michael is twenty-five.*

Michael:
1. *I am younger than George.*
2. *George is twenty-three.*
3. *George is three years younger than Harry.*

To make matters even more puzzling for the interviewer, only two of the three statements that each man made were true. Can you figure out the age of each one?

Answers and Solutions

6 **Answer:** *Jim robbed the bank.*

Solution: Start by setting up a simple truth chart similar to the one in example 1 above:

	STATEMENT	TRUTH VALUE
JIM	I'M INNOCENT.	
BUD	I'M INNOCENT.	
SAM	BUD IS THE GUILTY ONE.	

You are told that only one of the three statements turned out to be true. That means that 1T and 2F's have to be assigned to the chart in a logically consistent way.

Bud's statement—*I'm innocent*—and Sam's statement—*Bud is the guilty one*—contradict each other. So, one is true and the other false. Assume that Bud's statement is false and Sam's true:

	STATEMENT	TRUTH VALUE
JIM	I'M INNOCENT.	
BUD	I'M INNOCENT.	F
SAM	BUD IS THE GUILTY ONE.	T

Under the condition of the puzzle—that there be 1T and 2F's in the chart—Jim's statement must necessarily be assigned an F-value:

	STATEMENT	TRUTH VALUE
JIM	I'M INNOCENT.	F
BUD	I'M INNOCENT.	F
SAM	BUD IS THE GUILTY ONE.	T

Now, consider the logical outcome of this arrangement of 1T and 2F's. Jim's statement has an F-value. So, he was the guilty party—the opposite of *I'm innocent* is *I'm guilty.* Bud said that he, too, was innocent. According to the chart, his statement is also false. So, Bud was the guilty party as well. But there was only one robber. Obviously, since the above arrangement of 1T and 2F's leads to a contradiction, it must be rejected. However, in the process you have learned that Bud's statement is true and Sam's false—the reverse of your initial assumption:

	STATEMENT	TRUTH VALUE
JIM	I'M INNOCENT.	
BUD	I'M INNOCENT.	T
SAM	BUD IS THE GUILTY ONE.	F

Because there was only one true statement, Jim's statement turns out to be false again:

	STATEMENT	TRUTH VALUE
JIM	I'M INNOCENT.	F
BUD	I'M INNOCENT.	T
SAM	BUD IS THE GUILTY ONE.	F

Since Jim's statement—*I'm innocent*—is false, it can be concluded that he was, after all, the robber. Bud's statement—*I'm innocent*—can now be seen to be true, as correctly indicated by the T-value opposite his statement. And Sam's statement—*Bud is the guilty one*—can be seen to be false, as is indicated by the F-value opposite his statement.

7 **Answer:** *The young man belonged to the lie-telling clan.*

Solution: This puzzle is a version of example 2 above. The key to its solution is the fact that the answer to the question *Are you a truth-teller?* would always be *yes,* no matter who answered it.

Assume that the female informant was a truth-teller. Then, in response to the question, *Are you a truth-teller?* she would obviously have answered *yes* in her native tongue, for that would indeed have been the truth:

Question: *Are you a truth-teller?*
Truth-teller: *Ooma tooma* = *Yes.*
(which is true)

If, instead, she belonged to the lie-telling clan, then she would also have responded *yes* in her native tongue, because that answer would have been a lie in accordance with her mendacious nature. She does not, contrary to what she says, belong to the truth-telling clan.

Question: *Are you a truth-teller?*
Lie-teller: *Ooma tooma* = *Yes.*
(which is false)

So, the young man obviously lied when he said that *She said no.* Obviously, he belonged to the lie-telling clan.

8 **Answer:** *The male belonged to the truth-telling clan and the female to the lie-telling clan.*

Solution: This is a slightly trickier version of the previous puzzle. But the same line of reasoning applies here as well. Once again, the key to its solution is the fact that the female's answer to the question *Are you a truth-teller?* would have been *yes,* no matter to which clan she actually belonged.

Assume that the female informant was indeed a truth-teller. Then, in response to the question, *Are you a truth-teller?* she would obviously have answered *yes* in her dialect, for that would indeed have been the truth:

Question: *Are you a truth-teller?*
Truth-teller: *Goo boo* = *Yes.*
(which is true)

If, instead she belonged to the lie-telling clan, then she would also have responded *yes* in her dialect, because that answer would have been a lie, in accordance with her mendacious nature:

Question: *Are you a truth-teller?*
Lie-teller: *Goo boo* = *Yes.*
(which is false)

Now, analyze what the male informant said in its two parts:

Male Informant: *She said yes.*

Regardless of the clan to which the female informant actually belonged, the male's statement was obviously true, because, as was just demonstrated, the female had in fact answered *yes*. So, he told the truth and is, therefore, a truth-teller. Therefore, the next part of his statement is necessarily true:

Male Informant: *But she is a liar.*

So, the female lied in reality, and therefore belonged to the lie-telling clan.

9 **Answer:** *It is Friday.*

Solution: Start by setting up a truth chart:

	STATEMENT	TRUTH VALUE
ANITA	THE DAY BEFORE YESTERDAY WAS FRIDAY OR SATURDAY.	
BRUNHILDE	NO, YOU'RE WRONG. TODAY IS SATURDAY.	
CLAUDIA	NO, TODAY IS NOT SATURDAY. NOR IS IT SUNDAY OR MONDAY.	
DANIELA	THE DAY AFTER TOMORROW IS TUESDAY.	
ELLA	TOMORROW CANNOT BE FRIDAY OR SATURDAY.	
FRIDA	TOMORROW IS FRIDAY.	

Brunhilde's statement—*Today is Saturday*—contradicts Claudia's statement—*No, today is not Saturday.* One must therefore be true and the other false. Assume that Brunhilde's statement is the true one and Claudia's the false one. Since you are told that there is only one true statement, you can now fill in the chart with a T-value opposite Brunhilde's statement and with F-values opposite all the other statements:

	STATEMENT	TRUTH VALUE
ANITA	THE DAY BEFORE YESTERDAY WAS FRIDAY OR SATURDAY.	F
BRUNHILDE	NO, YOU'RE WRONG. TODAY IS SATURDAY.	T
CLAUDIA	NO, TODAY IS NOT SATURDAY. NOR IS IT SUNDAY OR MONDAY.	F
DANIELA	THE DAY AFTER TOMORROW IS TUESDAY.	F
ELLA	TOMORROW CANNOT BE FRIDAY OR SATURDAY.	F
FRIDA	TOMORROW IS FRIDAY.	F

Now, consider the logical outcome of assigning 1T and 5F's in this way. From Brunhilde's statement, it can be deduced that today is Saturday. However, when you consider Claudia's statement in its entirety—*No, today is not Saturday. Nor is it Sunday or Monday*—a contradiction emerges. The chart shows an F-value opposite her statement. And, indeed, the first part of her statement—*Today is not Saturday*—is manifestly false. But the second part of her statement—*Nor is it Sunday or Monday*—turns out to be true. Today (=Saturday), in fact, is neither Sunday nor Monday—just as Claudia asserts.

Since the initial assumption has led to a contradiction, it must be rejected. However, in the process you have learned that Claudia's statement is the true one and that Brunhilde's is one of the five false ones:

	Statement	Truth Value
Anita	The day before yesterday was Friday or Saturday.	F
Brunhilde	No, you're wrong. Today is Saturday.	F
Claudia	No, today is not Saturday. Nor is it Sunday or Monday.	T
Daniela	The day after tomorrow is Tuesday.	F
Ella	Tomorrow cannot be Friday or Saturday.	F
Frida	Tomorrow is Friday.	F

From Claudia's statement, you can now establish for certain that today is not Saturday, Sunday, or Monday. Now, consider Anita's statement—*The day before yesterday was Friday or Saturday*—which really consists of two assertions:

1. *The day before yesterday was Friday.* Therefore, today is Sunday:

Day before yesterday	Yesterday	Today
↓	↓	↓
Friday	*Saturday*	*Sunday*

2. *The day before yesterday was Saturday.* Therefore, today is Monday:

Day before yesterday	Yesterday	Today
↓	↓	↓
Saturday	*Sunday*	*Monday*

So, the gist of her statement is that today is either Sunday or Monday. But you have shown this to be false. Thus, the F-value in the truth chart opposite Anita's statement is the appropriate one. So far, so good.

You can skip over the next two statements, because you have already proven that Brunhilde's statement is false and Claudia's true. So, move on to Daniela's statement—*The day after tomorrow is Tuesday.* If the day after tomorrow were, in fact, Tuesday, as Daniela asserts, then today would be Sunday:

Today	Tomorrow	Day After Tomorrow
↓	↓	↓
Sunday	*Monday*	*Tuesday*

But you know that this is false. Thus, the F-value in the truth chart opposite Daniela's statement is also appropriate.

Next, consider Ella's statement—*Tomorrow cannot be Friday or Saturday.* Her statement really consists of two assertions:

1. *Tomorrow is not Friday.* So, today is not Thursday.
2. *Tomorrow is not Saturday.* So, today is not Friday.

In effect, Ella is saying that today cannot be Thursday or Friday, but her statement shows an F-value. Therefore, the opposite of what she tells us is true—namely, that it is indeed Thursday or Friday. To put it another way, if it is false that it is not Thursday or Friday, then it is Thursday or Friday!

Now, consider Frida's statement—*Tomorrow is Friday.* This would mean, of course, that today is Thursday. But, according to the truth chart, her statement has an F-value. So, contrary to what Frida says, today is not Thursday. From Ella's statement above, you have established that today is either Thursday or Friday. So, if it is not Thursday, then by elimination it follows that today is Friday. This is, in fact, the solution to the puzzle.

10 **Answer:** *Michael is twenty-two, George twenty-three, and Harry twenty-five.*

Solution: As usual, the first thing to do is to set up an appropriate truth chart:

	STATEMENT	TRUTH VALUE
GEORGE	1. I'M TWENTY-TWO. 2. HARRY IS TWO YEARS OLDER THAN I AM. 3. MICHAEL IS ONE YEAR YOUNGER THAN I AM.	1. 2. 3.
HARRY	1. I AM NOT THE YOUNGEST. 2. THERE IS A THREE-YEAR DIFFERENCE BETWEEN MY AGE AND MICHAEL'S AGE. 3. MICHAEL IS TWENTY-FIVE.	1. 2. 3.
MICHAEL	1. I AM YOUNGER THAN GEORGE. 2. GEORGE IS TWENTY-THREE. 3. GEORGE IS THREE YEARS YOUNGER THAN HARRY.	1. 2. 3.

Since each man made two true statements, your task is to put 2T's and 1F opposite each man's three statements in a logically consistent way. You can immediately see that George's first statement—*I'm twenty-two*—contradicts Michael's second statement—*George is twenty-three.* So, one is true and the other false. Assume that George's statement is true and Michael's false:

	STATEMENT	TRUTH VALUE
GEORGE	1. I'M TWENTY-TWO. 2. HARRY IS TWO YEARS OLDER THAN I AM. 3. MICHAEL IS ONE YEAR YOUNGER THAN I AM.	1. T 2. 3.
HARRY	1. I AM NOT THE YOUNGEST. 2. THERE IS A THREE-YEAR DIFFERENCE BETWEEN MY AGE AND MICHAEL'S AGE. 3. MICHAEL IS TWENTY-FIVE.	1. 2. 3.
MICHAEL	1. I AM YOUNGER THAN GEORGE. 2. GEORGE IS TWENTY-THREE. 3. GEORGE IS THREE YEARS YOUNGER THAN HARRY.	1. 2. F 3.

You can now complete Michael's set of truth-values with 2T's:

	STATEMENT	TRUTH VALUE
GEORGE	1. I'M TWENTY-TWO. 2. HARRY IS TWO YEARS OLDER THAN I AM. 3. MICHAEL IS ONE YEAR YOUNGER THAN I AM.	1. T 2. 3.
HARRY	1. I AM NOT THE YOUNGEST. 2. THERE IS A THREE-YEAR DIFFERENCE BETWEEN MY AGE AND MICHAEL'S AGE. 3. MICHAEL IS TWENTY-FIVE.	1. 2. 3.
MICHAEL	1. I AM YOUNGER THAN GEORGE. 2. GEORGE IS TWENTY-THREE. 3. GEORGE IS THREE YEARS YOUNGER THAN HARRY.	1. T 2. F 3. T

According to Michael's third statement, which has a T-value, George is three years younger than Harry. That is the same as saying that *Harry is three years older than George.* So, George's second statement—*Harry is two years older than I am*—is clearly false. Harry cannot be both two years and three years older than George:

	Statement	Truth Value
George	1. I'm twenty-two.	1. T
	2. Harry is two years older than I am.	2. F
	3. Michael is one year younger than I am.	3.
Harry	1. I am not the youngest.	1.
	2. There is a three-year difference between my age and Michael's age.	2.
	3. Michael is twenty-five.	3.
Michael	1. I am younger than George.	1. T
	2. George is twenty-three.	2. F
	3. George is three years younger than Harry.	3. T

As you can now see, George's third statement is to be assigned a T-value:

	Statement	Truth Value
George	1. I'm twenty-two.	1. T
	2. Harry is two years older than I am.	2. F
	3. Michael is one year younger than I am.	3. T
Harry	1. I am not the youngest.	1.
	2. There is a three-year difference between my age and Michael's age.	2.
	3. Michael is twenty-five.	3.
Michael	1. I am younger than George.	1. T
	2. George is twenty-three.	2. F
	3. George is three years younger than Harry.	3. T

According to George's first statement, which has a T-value, he is twenty-two years old; and according to his third statement, which also has a T-value, Michael is one year younger (twenty-one). Keep these two facts in mind as you go forward:

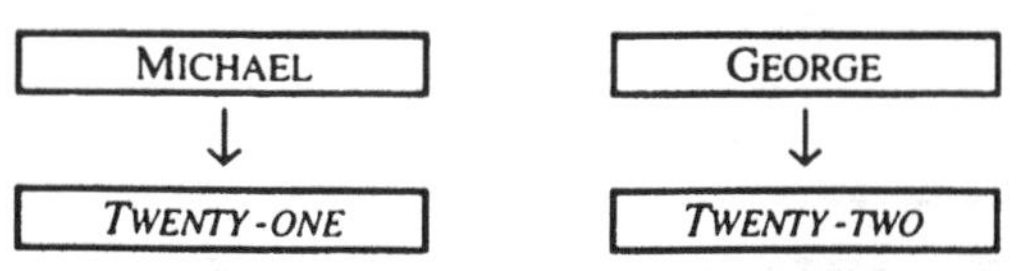

Now, it can be seen that Harry's third statement—*Michael is twenty-five*—is his false one. Therefore, his other two statements are to be assigned T-values:

	Statement	Truth Value
George	1. I'm twenty-two.	1. T
	2. Harry is two years older than I am.	2. F
	3. Michael is one year younger than I am.	3. T
Harry	1. I am not the youngest.	1. T
	2. There is a three-year difference between my age and Michael's age.	2. T
	3. Michael is twenty-five.	3. F
Michael	1. I am younger than George.	1. T
	2. George is twenty-three.	2. F
	3. George is three years younger than Harry.	3. T

Consider Harry's first and second statements, which have T-values, together: (1) *I am not the youngest;* (2) *There is a three-year difference between my age and Michael's age.* You have tentatively established that Michael is twenty-one and younger than George at twenty-two. So, Harry cannot be *three years younger* than Michael, for that would make him the youngest of the three. So, Harry is three years older, or twenty-four:

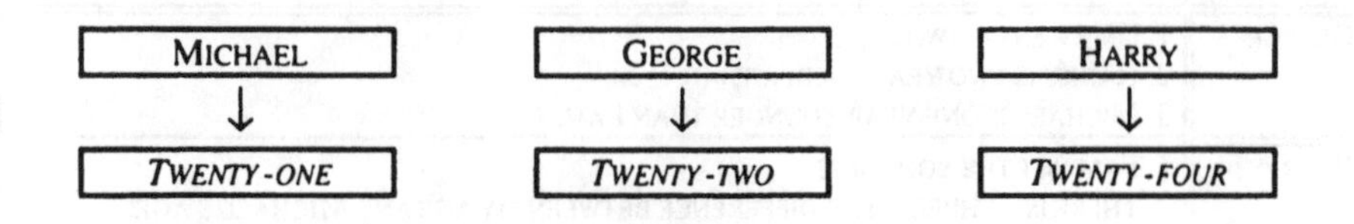

Now, reconsider George's second statement—*Harry is two years older than I am.* Looking at the age summaries above, you can see that this statement is true. But, according to the chart, it has an F-value. So, you have run up against a contradiction and must therefore discard your initial assumption—that George's first statement is true and Michael's second statement is false. But, in the process, you have discovered that the opposite holds. So, start over, noting this finding in your chart:

	STATEMENT	TRUTH VALUE
GEORGE	1. I'M TWENTY-TWO. 2. HARRY IS TWO YEARS OLDER THAN I AM. 3. MICHAEL IS ONE YEAR YOUNGER THAN I AM.	1. F 2. 3.
HARRY	1. I AM NOT THE YOUNGEST. 2. THERE IS A THREE-YEAR DIFFERENCE BETWEEN MY AGE AND MICHAEL'S AGE. 3. MICHAEL IS TWENTY-FIVE.	1. 2. 3.
MICHAEL	1. I AM YOUNGER THAN GEORGE. 2. GEORGE IS TWENTY-THREE. 3. GEORGE IS THREE YEARS YOUNGER THAN HARRY.	1. 2. T 3.

You can now complete George's set of statements with 2T's:

	STATEMENT	TRUTH VALUE
GEORGE	1. I'M TWENTY-TWO. 2. HARRY IS TWO YEARS OLDER THAN I AM. 3. MICHAEL IS ONE YEAR YOUNGER THAN I AM.	1. F 2. T 3. T
HARRY	1. I AM NOT THE YOUNGEST. 2. THERE IS A THREE-YEAR DIFFERENCE BETWEEN MY AGE AND MICHAEL'S AGE. 3. MICHAEL IS TWENTY-FIVE.	1. 2. 3.
MICHAEL	1. I AM YOUNGER THAN GEORGE. 2. GEORGE IS TWENTY-THREE. 3. GEORGE IS THREE YEARS YOUNGER THAN HARRY.	1. 2. T 3.

From George's second statement, which has a T-value, you can establish that Harry is two years older than he is. Another way of putting it is to say that *George is two years younger than Harry.* So, Michael's third statement—*George is three years younger than Harry*—can now be seen to be false. This makes his first statement necessarily true according to the condition that there be 2T's and 1F per set of statements:

	Statement	Truth value
George	1. I'm twenty-two. 2. Harry is two years older than I am. 3. Michael is one year younger than I am.	1. F 2. T 3. T
Harry	1. I am not the youngest. 2. There is a three-year difference between my age and Michael's age. 3. Michael is twenty-five.	1. 2. 3.
Michael	1. I am younger than George. 2. George is twenty-three. 3. George is three years younger than Harry.	1. T 2. T 3. F

From George's truthful second statement, you now know that Harry is older than George, or to put it another way, that *George is younger than Harry.* From Michael's truthful first statement, you also know that Michael is younger than George. So, Michael is the youngest of the three. Therefore, Harry's statement that he is not the youngest is true:

	Statement	Truth value
George	1. I'm twenty-two. 2. Harry is two years older than I am. 3. Michael is one year younger than I am.	1. F 2. T 3. T
Harry	1. I am not the youngest. 2. There is a three-year difference between my age and Michael's age. 3. Michael is twenty-five.	1. T 2. 3.
Michael	1. I am younger than George. 2. George is twenty-three. 3. George is three years younger than Harry.	1. T 2. T 3. F

From Michael's second statement, which has a T-value, you can establish that George is twenty-three. Since Michael is younger than George, then he clearly cannot be twenty-five, as Harry claims he is. So, Harry's third statement is false, leaving his second statement as being necessarily true:

	Statement	Truth value
George	1. I'm twenty-two. 2. Harry is two years older than I am. 3. Michael is one year younger than I am.	1. F 2. T 3. T
Harry	1. I am not the youngest. 2. There is a three-year difference between my age and Michael's age. 3. Michael is twenty-five.	1. T 2. T 3. F
Michael	1. I am younger than George. 2. George is twenty-three. 3. George is three years younger than Harry.	1. T 2. T 3. F

Now, you can establish the ages of the three. You know that George is twenty-three from Michael's second statement. George's third statement tells you that Michael is one year younger. So, Michael is twenty-two. And George's second statement tells you that Harry is two years older than George is. So, Harry is twenty-five:

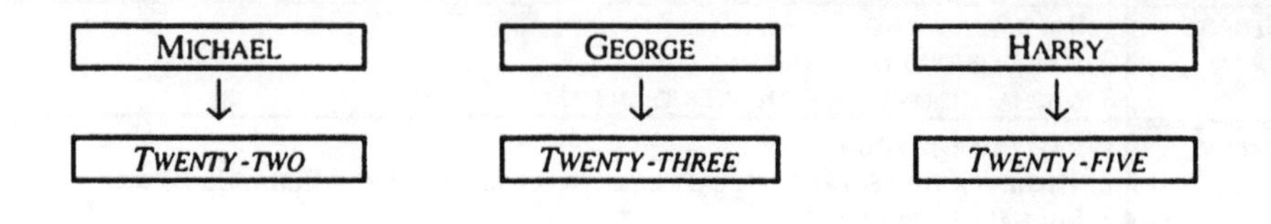

3 Puzzles in Trick Logic

So far, you have been honing your puzzle-solving skills in a systematic fashion by working with puzzles that involve straightforward reasoning. The first chapter gave you exposure to the mechanics of solving puzzles using deductive and inferential logic. The second chapter allowed you to become familiar with the trial-and-error method of analysis that solving truth logic puzzles entails.

The puzzles in this chapter belong to a whole different species of enigmas. Their solutions are not at all based on forthright logical reasoning. Rather, they play cunningly on deliberately deceptive statements, or else they involve an unexpected turn here and there. They will teach you, therefore, that it is always important to scrutinize the statements of all puzzles first and foremost for tricks, pitfalls, or logical twists.

How To...

As you might expect, there really is no sure way to detect the trickery concealed in puzzles of this type. You will just have to be constantly on your guard against ruses or decoys in either the *wording* or the *layout* of the puzzle.

Puzzle Properties

There are three main types of trick puzzles to be on guard against: (1) puzzles that play on the meanings of specific words; (2) puzzles that entice you to work through a series of calculations that end up being totally worthless; and (3) puzzles that play on the layout of some argument or mathematical proof.

Example 1 Here is a problem that seems to involve a straightforward calculation, but beware of its hidden trap!

> How much dirt is there in a hole that is 1 foot wide by 1 foot long by 1 foot deep?

If your answer was $1 \times 1 \times 1 = 1$ cubic foot of dirt, you fell into a word trap. Obviously, there is no dirt in a *hole*. A *hole* has *nothing* in it, by definition! What you have in effect calculated is the number of cubic feet of dirt *taken out* of the ground. But the puzzle does not tell you to do that!

Example 2 The next one is a version of a word trick puzzle that is included regularly in puzzle books.

> In my right hand I have two current U.S. coins. The two coins add up to 15¢. One of the two coins is not a nickel. So, what are the two coins I have in my hand?

If you try out various combinations of two current U.S. coins, excluding a nickel, you will become frustrated. There simply is no pair of coins adding up to 15¢, unless one of them is a nickel. The trap in this case is in the puzzle's wording. The puzzle says that there are *two* coins that add up to 15¢, and that *one* of the two coins is not the 5¢ coin; so, the only conclusion that can safely be reached is that the *other* coin is the nickel. More specifically, *one* of the two coins is a 10¢ coin (a dime, not a nickel), while the *other* one is the nickel.

Summary

To increase your chances of detecting or fleshing out the concealed trap or twist in any puzzle, write out all puzzles for yourself, inspecting carefully what they say, word by word. There really is nothing else that can be done. With practice, you will develop a knack for spotting the trick in such puzzles. Be careful! Every once in a while, the trap is much larger than just isolated words; it might encompass the whole statement of the puzzle.

Good luck! You will really need it!

Puzzles 11–24

Answers, along with step-by-step solutions, can be found at the end of the chapter.

11 If it takes 3 minutes to boil an egg, how long would it take to boil three eggs?

12 Farmer Zack has three and seven-ninths haystacks in one part of his field, and he has two and two-thirds haystacks in another part. If he puts all his haystacks together, how many haystacks will he have?

Is it legal in California for a man to marry his widow's sister?

14 A bull is put on a weighing scale, but he is so big that only three of his four legs will fit on the scale. The scale shows 1,000 pounds. How much do you estimate the bull weighs when he stands on all four legs?

15 It takes twelve 1¢ stamps to make a dozen. How many 2¢ stamps does it take?

16 An explorer sets up camp. She walks 1 mile south from the camp, turns, and walks 1 mile east. Then, she turns again and walks 1 mile due north. At that point she finds herself back where she started! How is that possible?

17 After a series of experiments, a famous chemist discovered that it took 80 minutes for a specific chemical reaction to occur when he was wearing his glasses, but that it took the same reaction an hour and 20 minutes to occur when he was not wearing them. Why?

18 What is 30 divided by ½?

19 What two whole numbers, *not fractions,* make 13 when they are multiplied together?

20 Do you think that the product of the first ten digits is between 100 and 1,000, or is greater than 1,000?

21 Why are 1997 dollar bills worth more than 1980 dollar bills?

22 Two train tracks run parallel, except for a place in a tunnel where they merge into a single track. One day a train entered the tunnel going in one direction and another entered the tunnel going in the opposite direction. Both trains were moving at top speed, yet there was no collision. Why?

23 Three women decide to go on a holiday to Las Vegas. They share a room at a hotel that is charging 1920s rates as a promotional gimmick. The women are charged only $10 each, or $30 in all. After going through his guest list, the manager discovers that he has made a mistake and has actually overcharged the three vacationers. The room the three are in costs only $25. So, he gives a bellhop $5 to return to them. The sneaky bellhop knows that he cannot divide $5 into three equal amounts. Therefore, he pockets $2 for himself and returns only $1 to each woman.

Now, here's the conundrum. Each woman paid $10 originally and got back $1. So, in fact, each woman paid $9 for the room. The three of them together thus paid $9 × 3, or $27 in total. If we add this amount to the $2 that the bellhop had pocketed dishonestly, we get a total of $29. Yet the women paid out $30 originally! Where is the other dollar?

24 Yesterday, in a suburban mall, the first customer in a bookstore gave the salesclerk a $10 bill for a $3 book. The salesclerk, having no change, took the $10 bill across the corridor to the record store to get it broken down into ten $1 bills. The salesclerk then gave the customer the book worth $3 and seven $1 bills as change.

An hour later, the record-store salesclerk brought back the $10 bill demanding her money back from the bookstore salesclerk, claiming that the bill was counterfeit. To avoid quarreling, the bookstore salesclerk decided to give her ten $1 bills, taking back the counterfeit. That means that the bookstore salesclerk was out $3 (=cost of the book), plus the ten $1 bills he gave to the record-store salesclerk. Altogether he lost $13. But only $10 were used in the whole transaction. What happened?

Answers and Solutions

11 **Answer:** *Three minutes.*

Solution: Obviously, you would not boil 3 eggs separately, one after the other. If you did, you would, of course, need 3 minutes to boil the first egg, then another 3 minutes to boil the second egg, and then again another 3 minutes to boil the third egg—or 9 minutes in total. But, if you boil one, two, three, or a million eggs *at one time,* then it will take only 3 minutes.

If you still do not see this, do the following: (1) Put one egg in a pot and boil it for 3 minutes. (2) Put two eggs in a pot and boil them together. How long did it take? Just like before, of course, 3 minutes (because the two eggs are boiling together in the same pot). (3) Put 3 eggs in a pot and boil them together. How long did it take? Three minutes.

12 **Answer:** *One huge haystack.*

Solution: If you do not see this, conduct the following experiment with feathers. In one part of your living room, make three and seven-ninths piles of feathers (approximately, of course). In another part of the room, make two and two-thirds piles of feathers (again, approximately). Now put all the piles together. How many piles of feathers do you now have?

13 **Answer:** *The situation is impossible.*

Solution: The trick in this puzzle is the word *widow.* A man who leaves a *widow* is a dead man, of course! So, how can a dead man marry his widow's sister?

14 **Answer:** *1,000 pounds.*

Solution: The bull weighs 1,000 pounds, no matter if he stands on four legs or on three. If you do not see this, perform the following analogous experiment. Stand on a weighing scale. Look at the weight that the scale shows. Now, lift up one of your feet, staying on the scale. Has your weight changed? Of course it hasn't. See the point?

15 **Answer:** *Twelve.*

Solution: The price of a stamp means nothing. To make a *dozen* stamps, you will need 12 stamps. These can be 1¢ each (i.e., 12 stamps costing 1¢ each will make a dozen); these can be 2¢ each (i.e., 12 stamps costing 2¢ each will also make a dozen); these can be 3¢ each (i.e., 12 stamps costing 3¢ each will likewise make a dozen); and so on.

16 **Answer:** *The explorer set up camp at the North Pole.*

Solution: If you drew a diagram to help you trace the explorer's path, you probably came up with something like this:

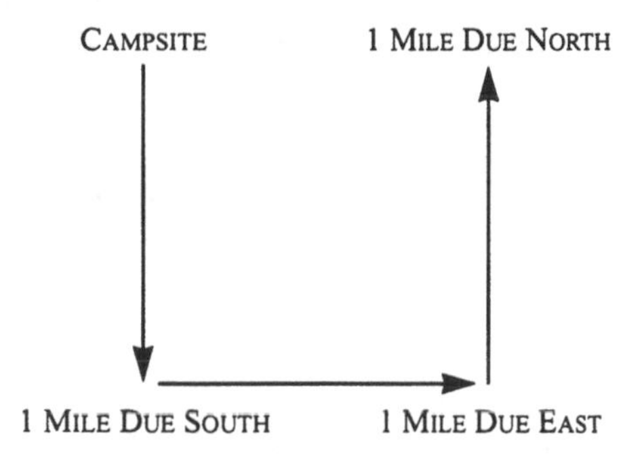

Clearly, according to this diagram, drawn on a two-dimensional sheet of paper (known as a plane in geometry), the explorer should not end up back at the campsite, as the puzzle states! But, then, what if the diagram were drawn on a sphere? That is, after all, the shape of the earth. The situation described in the puzzle would then make sense if the explorer pitched camp at the North Pole. From that location she could go south a mile, walk east, and then walk north, and she would be right back where she started from! If you do not see this, get a hold of a globe and try it out for yourself.

17 **Answer:** *1 hour and 20 minutes = 80 minutes.*

Solution: Since *1 hour* is equal to 60 minutes, then *1 hour and 20 minutes* is, of course, equal to 60 + 20 = 80 minutes. So, there is nothing to explain, because it took the reaction 1 hour and 20 minutes, or in equivalent terms, 80 minutes, to occur, no matter what the chemist was wearing.

18 **Answer:** *60*

Solution: The trick in this puzzle is the statement *30 divided by ½.* If you took ½ of 30 = 15, you forgot that dividing 30 by ½ is equivalent to multiplying it by 2.

If you do not see this, rephrase the question: *What is 30 divided by ½?* as *How many times does ½ go into 30?* Before answering this question, consider another question first: *How many times does 1 go into 30?* Obviously, 30 times. How much bigger is 1 than ½? It is, of course, twice as big (one pie is twice half a pie). So, dividing 30 by ½ means that you will get a result that is twice as big, namely $30 \times 2 = 60$.

19 **Answer:** *$13 \times 1 = 13$.*

Solution: The only two whole numbers that make 13 when multiplied together are 13 itself and 1. Incidentally, this is true of any *prime number:* the only two whole numbers that make any *prime* (3, 5, 7, 11, 13, 17, . . .) when multiplied together are the prime number itself and 1.

20 **Answer:** *The product is 0.*

Solution: The product of the first ten digits is 0, because the first ten digits—0, 1, 2, 3, 4, 5, 6, 7, 8, 9—include 0 among them. Any number or sequence of numbers multiplied by 0 will equal 0: $1 \times 0 = 0$; $1 \times 2 \times 0 = 0$; $1 \times 2 \times 3 \times 0 = 0$; and so on.

21 **Answer:** *One thousand nine hundred and ninety-seven dollars is obviously worth more than one thousand nine hundred and eighty dollars.*

Solution: If this puzzle stumped you, it is because you interpreted the digits 1997 and 1980 as representing calendar years. The puzzle obviously wanted you to interpret them literally as digits: *one thousand nine hundred and ninety-seven* dollars (=$1997) is obviously worth more than *one thousand nine hundred and eighty* dollars (=$1980).

22 **Answer:** *The trains entered the tunnel at different times.*

Solution: If this puzzle stumped you, it is because you made an unwarranted assumption. You assumed that the two trains entered the tunnel at the same time. Obviously, they did not. One train entered earlier than did the other.

23 **Answer:** *The women paid $27 dollars, of which the hotel got $25 and the bellhop $2.*

Solution: This is a classic in the genre of trick puzzles. Actually, the deception here is not in any single word, but in the way the puzzle lays out the numerical facts. Here's how they should have been laid out in order to avoid the apparent discrepancy.

Originally, the women paid out $30 for the room. That's how much money was in the hands of the hotel manager, when he realized that he had overcharged them. He kept $25 of the $30 and gave $5 to the bellhop to return to the women.

Now, focus your attention on the women. They each got back $1. This means, in effect, that they had paid $9 each for the room. Thus, altogether they paid out $27, which is $2 more than they should have paid for the room—namely, $25. As you know, these $2 were the ones pilfered by our devious bellhop!

In sum, there is no missing dollar. The women paid $27, of which the hotel got $25 and the bellhop $2.

24 **Answer:** *The bookstore salesclerk was out the $3 book and $7 from his pocket—$10 in total.*

Solution: This is another classic puzzle in trick logic. Like the previous puzzle, the deception here is not to be found in any specific word, but in the layout of the numerical facts. Here's how they should have been laid out in order to avoid the apparent discrepancy:

First, the bookstore salesclerk received nothing for the $3 book, since the counterfeit $10 bill was worth nothing. So, from the outset, he was out $3. That $3 went to the customer.

Now, consider what happened in the other transaction—the one between the bookstore salesclerk and record-store salesclerk. The bookstore salesclerk received ten genuine $1 bills. So, at first it was the record-store salesclerk who was out $10. When the bookstore salesclerk got back to his store, he gave $7 of the ten good bills to the customer, and put the remaining good $3 in his cash register. The end result of this transaction was that the bookstore salesclerk was out $7, which he gave to the customer. Altogether, the customer gained $10—a $3 book and $7 in good bills. That ends the bookstore salesclerk's transaction with the customer.

At this point, the bookstore salesclerk was out the $3 for the book, not the $7 that he gave back as change to the customer—that came out of the register of the record-store salesclerk. When the record-store salesclerk asked for her $10 back, the bookstore salesclerk still had the $3 in his pocket left over from the $10 she had given him previously—the other $7 went to the customer. So, he gave her back her $3 and made up the $7 difference from his own pocket. In total, therefore, the bookstore salesclerk was out the $3 book and the $7 from his pocket—$10 in total.

4 Puzzles in Arithmetical Logic

Arithmetic is, literally, the art of counting. In its oldest Greek form, the word comes from *arithmetike,* which combines the ideas of two words in Greek, *arithmos* "number" and *techne* "art, skill." Counting probably began before alphabetic writing. The cuneiform (baked-clay) tablets of the Sumerians and Babylonians about 5,000 years ago prove that even the earliest civilizations had sophisticated number systems for carrying out common business transactions, for measurement, and for many other practical activities.

Counting, number properties, and the basic operations of arithmetic—adding, subtracting, multiplying, and dividing—have also been the source of a myriad of puzzles throughout the centuries. As is the case with other kinds of puzzles, success at solving arithmetical puzzles depends largely on knowing how to attack them methodically and how to construct appropriate charts or diagrams.

You might be interested to know that the great German mathematician Karl Friedrich Gauss (1777–1855) was only 10 years old when he dazzled his teachers with his wizardry at arithmetical calculation. One day, his class was asked to cast the sum of all the numbers from one to one hundred: $1 + 2 + 3 + 4 + \ldots + 100 = ?$ Amazingly, Gauss raised his hand within seconds and gave the correct response of 5,050, while the other students continued to toil over this seemingly gargantuan arithmetical task. When his teacher asked him how he was able to come up with the answer so quickly, he replied as follows:

There are 49 pairs of numbers between one and one hundred that add up to one hundred: $1 + 99 = 100$, $2 + 98 = 100$, $3 + 97 = 100$, and so on. That makes 4,900, of course! The number 50, being in the middle, stands alone, as does 100, standing at the end. Adding 50 and 100 to 4,900 gives 5,050!

How To...

There are many types of puzzles in arithmetical logic, but the solutions to such puzzles invariably entail a few general lines of attack, two of which are illustrated below in examples 1 and 2. Basically, these puzzles require that you count correctly, make no assumptions, and, above all else, take no shortcuts. More often than not, you will have to go painstakingly through the step-by-step procedures involved in counting, measuring, or estimating something.

Puzzle Properties

You can always recognize an arithmetical puzzle by the fact that it involves counting, calculating, or executing an arithmetical operation. For example, you might be asked to calculate how many coins of a certain type will be needed to pay for something, how long it will take a car to pass another car going at half speed, how many weighings on a balance scale will be required to identify a billiard ball that weighs less than other billiard balls, how many days a snail requires to reach the top of a well if it both goes up and slides down a given number of feet each day, and so on.

Example 1 Here is a version of a classic puzzle in this genre.

> A snail is at the bottom of a 30-foot well. Each day it crawls up 3 feet and slips back 2 feet. At that rate, when will the snail be able to reach the top of the well?

Newcomers to this kind of puzzle have difficulties coming up with the correct answer. Since the snail crawls up 3 feet, but slips back 2 feet, its net distance gain at the end of every day is, of course, *1 foot up* from the day before. To put it another way, the snail's climbing rate is *1 foot up per day.*

At the end of the first day, therefore, the snail will have gone up 1 foot from the bottom of the well, and will have 29 feet left to go to the top (remember that the well is 30 feet deep). To keep track of the snail's daily progress, it is useful to draw a *flow chart* such as the following one:

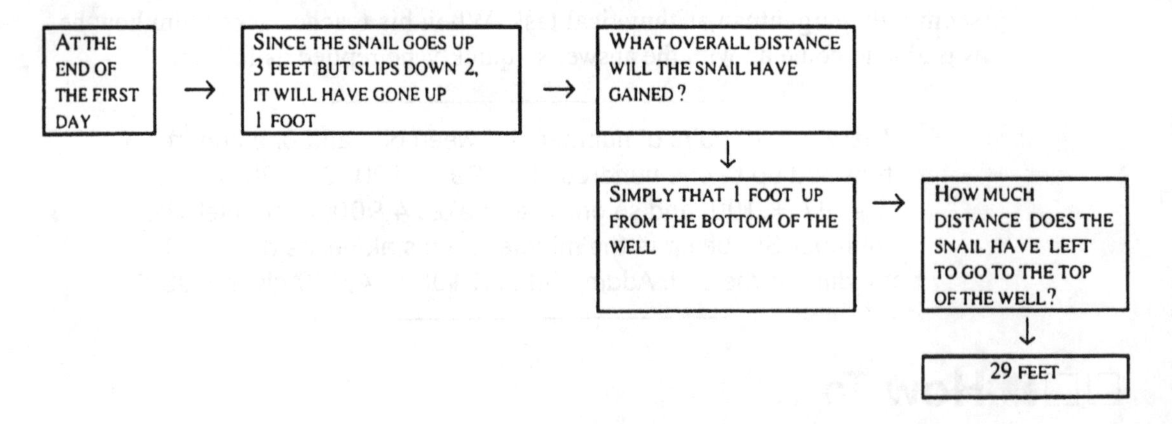

From this point on, you can conveniently record all your calculations directly on the flow chart. Here's how to record the snail's progress at the end of the second day:

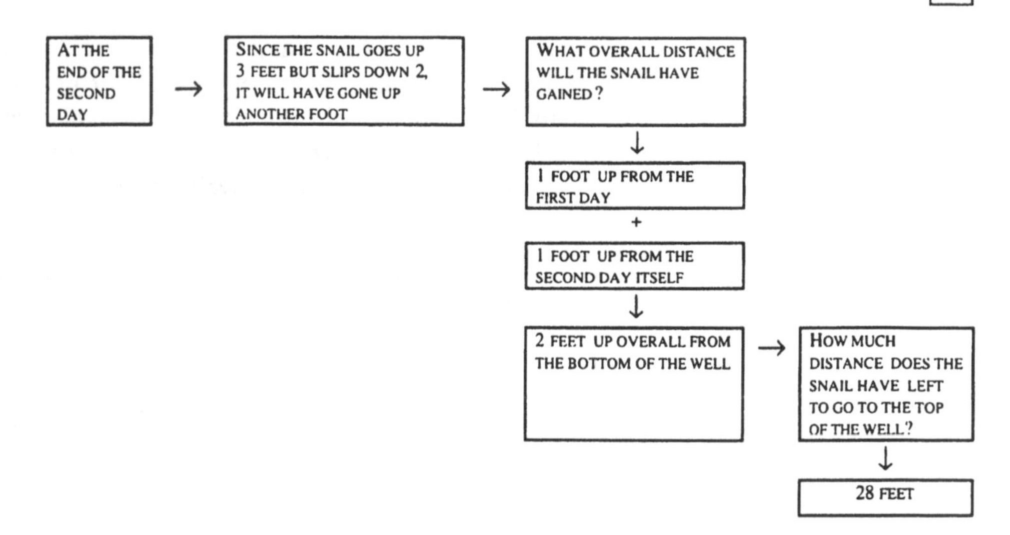

As you can see, at the end of the second day, the snail will have gone up 2 feet overall—*1 foot gained at the end of the first day + 1 foot gained at the end of the second day itself*—from the bottom of the well. You can see as well that it has 28 feet left to go to the top of the well. So, continuing on in this way, by the end of the third day, the snail will have gone up 3 feet overall from the bottom, and will have 27 feet to go to the top, by the end of the fourth, it will have gone up 4 feet overall, and will have 26 feet to go, and so on. However, be careful! If you conclude at this point that the snail will get to the top of the well on the 29th or 30th day, as many do, you will have fallen into a trap!

To see why, consider the snail's progress on the 27th day. At the end of that day the snail will have climbed up 27 feet overall from the bottom of the well, and it will have 3 feet left to go to the top:

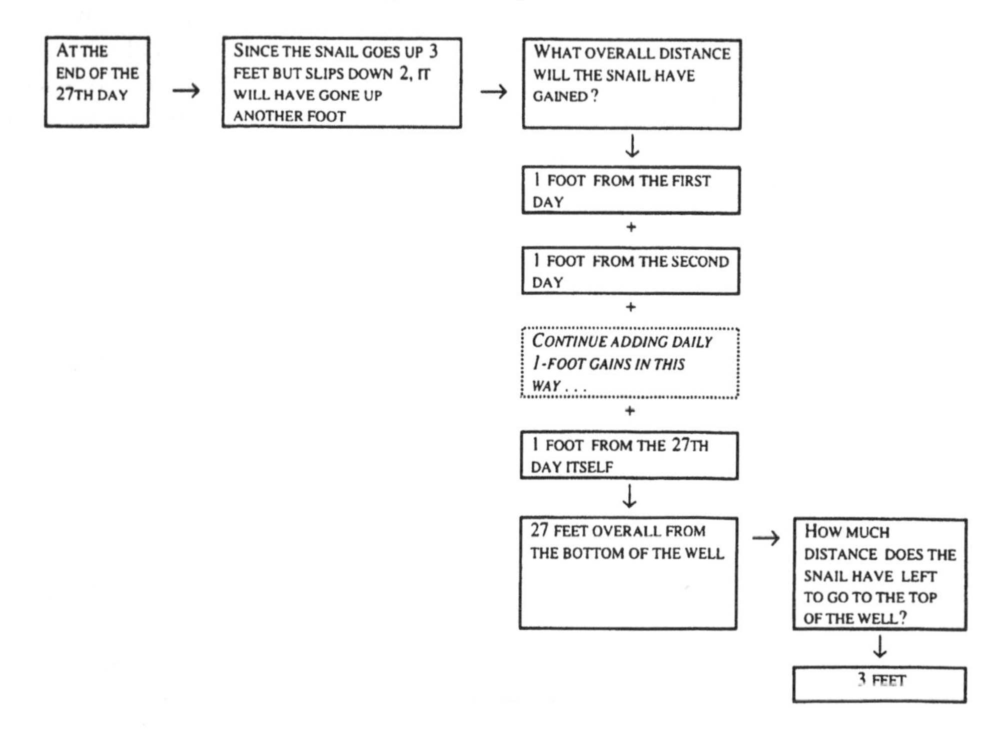

As you can see, on the morning of the 28th day the snail will need to go up just *3 more feet* to reach the top of the 30-foot well. So, during the *28th* day, how far up does the snail crawl before sliding down? As you know, the snail will go up 3 feet first, *before it starts sliding*. End of matter! Those 3 feet are enough to take the snail right up to the top of the well, where it can crawl out.

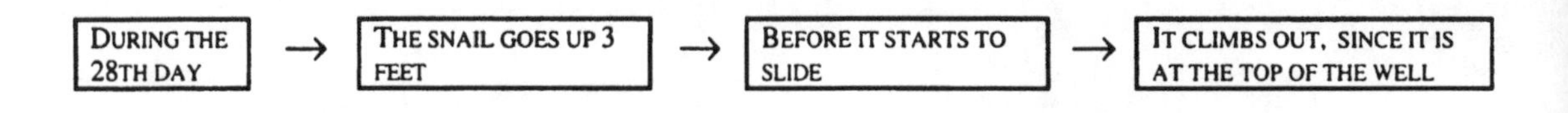
DURING THE 28TH DAY
→
THE SNAIL GOES UP 3 FEET
→
BEFORE IT STARTS TO SLIDE
→
IT CLIMBS OUT, SINCE IT IS AT THE TOP OF THE WELL

The solution is now complete. It takes the snail 28, not 29 or 30, days to climb to the top of the well.

Example 2 Like the previous one, the next puzzle also finds its way into most puzzle collections.

> I have six billiard balls, one of which weighs less than the other five. Otherwise, they all look exactly the same. How can I identify the one that weighs less on a balance scale with only two weighings?

Weighing puzzles such as this one are always best approached with trial runs using fewer items. These allow you to see if there is some general principle or procedure that can be applied to solving the original puzzle. For this particular puzzle, it is advisable to consider first the weighing of two balls. Clearly, you can put both balls on the scale pans at the same time—one on the left pan and one on the right pan. The pan that goes up, of course, is the one holding the ball that weighs less. In this case, one weighing was enough to identify the culprit ball.

Next, consider the weighing of four balls. First divide the four balls equally in half, that is, into two sets of two balls each.

Weighing 1 Put two balls on the left pan and two on the right pan this time. The pan that goes up contains the ball that weighs less, but you do not yet know which one of the two that it is.

Weighing 2 So, take the two suspect balls from the pan that went up, discarding the ones on the other pan. Put each one on a separate pan—one on the left pan and one on the right pan. The pan that goes up contains the ball that weighs less.

These two trial runs have shown you how to go about identifying the ball in a collection of any even number of balls that weighs less than the others. You are now ready to turn your attention to the original puzzle. There are six balls in the collection, and you are told to identify the culprit ball in only two weighings. Is that possible? Let's see. Start off in the same way as you did before: that is, divide the six balls equally in half, into two sets of three balls each. Then, go ahead and perform the first weighing as you also did before:

Weighing 1 Put three balls on each pan this time—three on the left pan and three on the right pan. The pan that goes up contains the ball that weighs less, but you do not yet know which one of the three.

Now, for the second weighing, focus your attention on the set containing the suspect ball, discarding the ones on the other pan. Is it possible to identify the culprit ball in just one more weighing (recall that the puzzle asks us to identify the culprit ball in just two weighings)? Well, you know how to weigh two of them easily. So, for the moment, select any two of the three balls to weigh, putting the third ball aside. Put each one of the two balls on a separate pan—one on the left pan and one on the right pan. What are the possible outcomes of this second weighing? Let's see, using a flow chart:

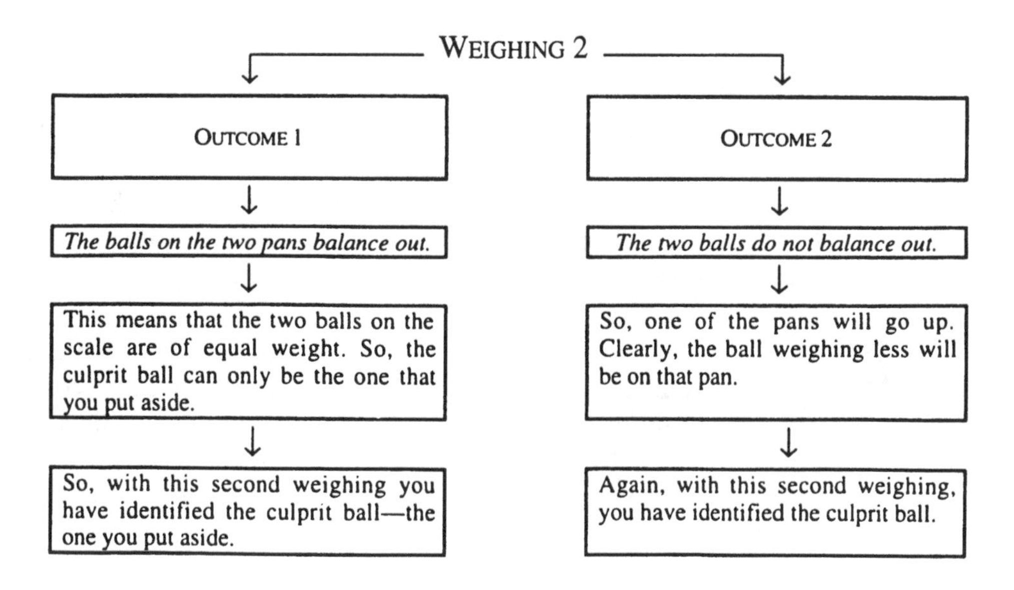

WEIGHING 2
OUTCOME 1
OUTCOME 2
The balls on the two pans balance out.
The two balls do not balance out.
This means that the two balls on the scale are of equal weight. So, the culprit ball can only be the one that you put aside.
So, one of the pans will go up. Clearly, the ball weighing less will be on that pan.
So, with this second weighing you have identified the culprit ball—the one you put aside.
Again, with this second weighing, you have identified the culprit ball.

In sum, it took just two weighings, and a little bit of crafty logical thinking, to identify the ball that weighs less.

Summary

Puzzles in arithmetical logic require that you make no assumptions in calculation and, more importantly, that you think through some calculation procedure—counting a snail's progress up a well on a day-by-day basis, measuring balls on a weighing scale—in a systematic way. Working out a solution for simpler versions of a puzzle is a basic strategy. This will allow you to see if there is a general procedure in the simpler versions that can be utilized profitably. So, for instance, if you are asked to weigh eight balls on two weighing pans using a specific number of weighings, you might start off by considering the weighing of two balls, then of four, and so on. When you detect a general weighing procedure, you will be ready to attack the original puzzle.

In summary, when solving a puzzle in arithmetical logic, remember to:

- Reduce it to its basic elements.
- Check out the calculation procedure you have used, in order to ascertain that you have made no erroneous assumptions or taken misleading shortcuts.
- Use flow charts and other kinds of diagrams to help you literally see the components of the calculation procedure in terms of their sequential organization.
- Go through some trial runs with less complex facts or figures, if called for, so that you can extrapolate any general principle or procedure that can be applied to the original puzzle.

Puzzles 25–34

Answers, along with step-by-step solutions, can be found at the end of the chapter.

25 I have seven billiard balls, one of which weighs less than the other six. Otherwise, they all look exactly the same. How can I identify the one that weighs less on a balance scale with no more than two weighings?

26 Jack used to smoke too much. One day he decided to quit, cold turkey, after smoking the 27 cigarettes that remained in his pocket. He took out the 27 cigarettes and started to smoke them, one by one. Since it was his habit to smoke only two-thirds of a cigarette, Jack soon realized that he could stick three butts together to make another cigarette. So, before giving up his bad habit, how many cigarettes did he end up smoking?

27 There are between 50 and 60 billiard balls in a box. If you count them 3 at a time, you will find that there are 2 left over. If, however, you count them 5 at a time, you will find that there are 4 left over. How many billiard balls are there in the box?

28 A train leaves New York for Washington every hour on the hour. A train leaves Washington for New York on the hour and on the half-hour. The trip takes 5 hours each way. If you are on the train from New York bound for Washington, how many of the trains coming from Washington and going toward New York would you pass?

29 There are two containers on a table, **A** and **B**. **A** is half full of wine, while **B**, which is twice **A**'s size, is one-quarter full of wine. Both containers are filled with water and the contents are poured into a third container, **C**. What portion of container **C**'s mixture is wine?

30 My friend Danielle is a mountain climber. She can hike uphill at an average rate of 2 miles per hour, and she can hike downhill at an average rate of 6 miles per hour. Going uphill and downhill, without staying at the top, what will her average speed for a whole trip be?

31 Jim has 20 coins in his pocket, consisting of dimes and nickels. Altogether, the coins add up to $1.35. How many of each does he have?

32 Two children, a boy and a girl, were out riding their bikes yesterday, coming at each other from opposite directions. When they were exactly 20 miles apart, they began racing toward each other. The instant they started, a fly on the handle of the girl's bike started flying toward the boy. As soon as it reached the handle on his bike, it turned and started back toward the girl. The fly flew back and forth in this way, from handlebar to handlebar, until the two bicycles met. Each bike moved at a constant speed of 10 miles an hour, and the swifter fly flew at a constant speed of 15 miles an hour. How much total distance did the fly cover?

33 Three books, each of the same width, are stacked upright against each other on a bookshelf. Each cover is ½-inch thick, and the width of the pages between the two covers is 2 inches. A bookworm starts on the first page of the first book to the left and bores its way straight through to the last page of the last book to the right. How far has the bookworm gone?

34 You are offered a new part-time job as a pizza delivery person, working only on weekends. Your boss gives you a choice in salary options as follows:

Option A $4,000 for your first year of work, and a raise of $800 for each year after the first.

Option B $2,000 for your first 6 months of work, and a raise of $200 every 6 months thereafter.

Which offer is the better one?

Answers and Solutions

25 **Answer:** *One or two weighings would do the job, as described below.*

Solution: This puzzle can be solved in exactly the same way in which example 2 above was solved. First, remove one of the seven balls and put it aside; then, put three balls on each pan—three on the left pan and three on the right pan. The following two scenarios are now imaginable:

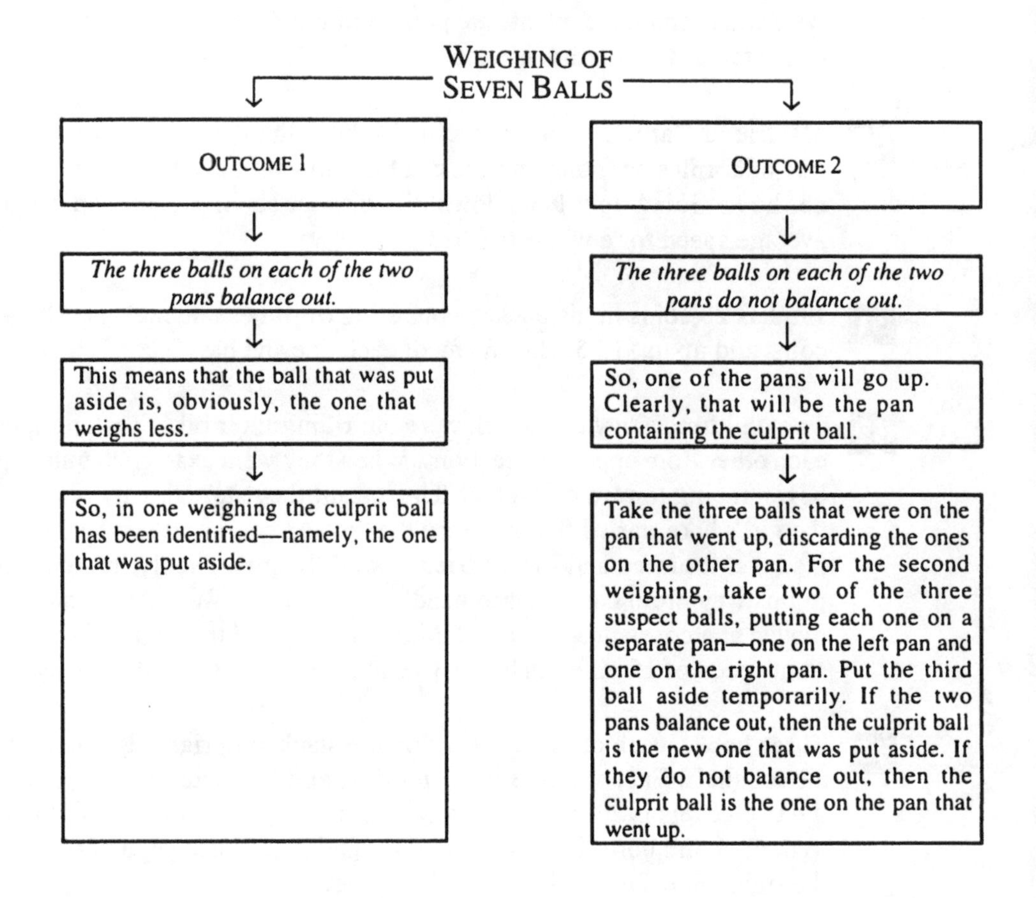

WEIGHING OF SEVEN BALLS
OUTCOME 1
OUTCOME 2
The three balls on each of the two pans balance out.
The three balls on each of the two pans do not balance out.
This means that the ball that was put aside is, obviously, the one that weighs less.
So, one of the pans will go up. Clearly, that will be the pan containing the culprit ball.
So, in one weighing the culprit ball has been identified—namely, the one that was put aside.
Take the three balls that were on the pan that went up, discarding the ones on the other pan. For the second weighing, take two of the three suspect balls, putting each one on a separate pan—one on the left pan and one on the right pan. Put the third ball aside temporarily. If the two pans balance out, then the culprit ball is the new one that was put aside. If they do not balance out, then the culprit ball is the one on the pan that went up.

In sum, scenario 1 took one weighing to carry out; scenario 2 took two weighings. So, it will take, at most, two weighings to identify the ball that weighs less than the others.

26 **Answer:** *40 cigarettes.*

Solution: You know that Jack smoked the 27 cigarettes he took out from his pocket. Since he smoked only two-thirds of a cigarette, he therefore would leave a butt equal to one-third of a cigarette. So, for every 3 cigarettes he smoked, he was able to piece together a new cigarette ($\frac{1}{3}$ butt + $\frac{1}{3}$ butt + $\frac{1}{3}$ butt = 1 new cigarette). After smoking the original 27 cigarettes, he was thus able to make 9 new cigarettes:

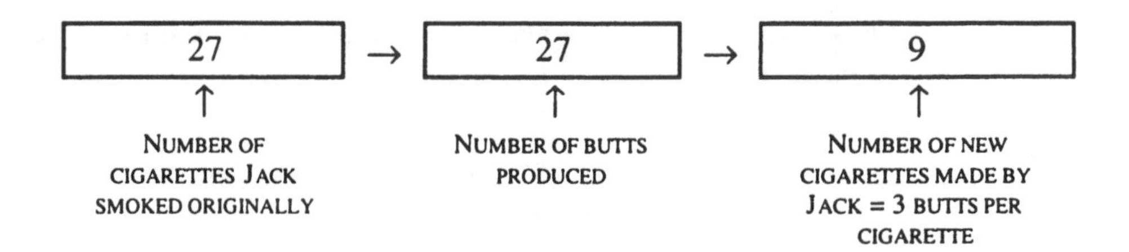

If you stopped here, simply adding 27 (number of cigarettes Jack smoked originally) + 9 (number of new cigarettes made and smoked by Jack) = 36 (total number of cigarettes smoked by Jack), you forgot that smoking the 9 new cigarettes also produced butts. In fact, Jack's 9 new cigarettes produced 9 new butts of their own. From these 9 butts, Jack was able to make, of course, 3 more cigarettes (3 butts = 1 new cigarette):

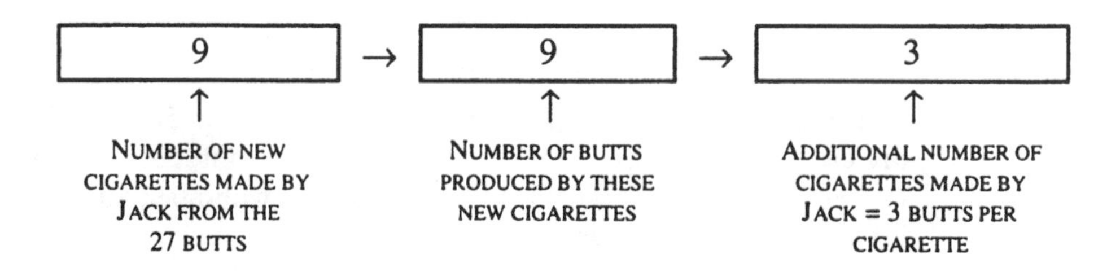

So, in addition to the 9 new cigarettes Jack made from the original 27, he was also able to make 3 more from those 9. But, then, those 3 extra cigarettes produced 3 butts of their own, from which Jack was able to make yet 1 more cigarette:

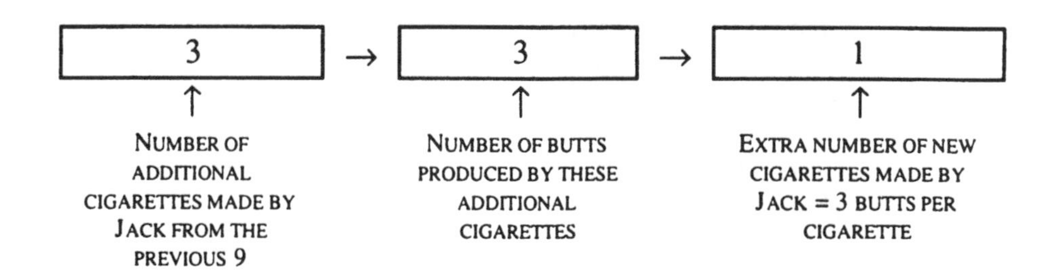

Altogether, therefore, Jack smoked 27 + 9 + 3 + 1 = 40 cigarettes before giving up his bad habit.

27 **Answer:** *59 balls.*

Solution: *Counting* a number of balls by ones, twos, threes, and so on is the equivalent of *dividing* that number of balls into smaller groups of one ball, two balls, three balls, and so on. So, to solve this puzzle, you must identify the number of balls between 50 and 60 which, when divided by 3, gives a remainder of 2—a remainder of 2 is equivalent to saying that there are 2 balls left over—and when divided by 5, gives a remainder of 4.

First, divide the numbers between 50 and 60 by 3, identifying those that leave a remainder of 2:

$50 \div 3 = 16$, remainder = 2
$51 \div 3 = 17$, no remainder
$52 \div 3 = 17$, remainder = 1
$53 \div 3 = 17$, remainder = 2
$54 \div 3 = 18$, no remainder
$55 \div 3 = 18$, remainder = 1
$56 \div 3 = 18$, remainder = 2
$57 \div 3 = 19$, no remainder
$58 \div 3 = 19$, remainder = 1
$59 \div 3 = 19$, remainder = 2
$60 \div 3 = 20$, no remainder

You can now see that the only candidates between 50 and 60, which when divided by 3 will leave a remainder of 2, are the numbers 50, 53, 56, and 59. Discard the others, proceeding to determine which number, 50, 53, 56, or 59 will leave a remainder of 4 when it is divided by 5:

$50 \div 5 = 10$, no remainder
$53 \div 5 = 10$, remainder = 3
$56 \div 5 = 11$, remainder = 1
$59 \div 5 = 11$, remainder = 4

As you can see, 59 balls is the answer. And, in fact, when you count 59 balls 3 at a time, you'll get 2 left over; but when you count them 5 at a time, you'll get 4 left over.

28 **Answer:** *10 trains.*

Solution: This puzzle shows why a diagram is such a useful tool for helping the solver literally visualize what a puzzle says. Let's say that you get on a train at the New York station at 12:00 noon. It could be at any other time; the reasoning will be the same.

As mentioned in the puzzle, five hours later, at 5:00 P.M., your train arrives at the Washington station. Now, draw a diagram charting the relative positions of the trains that were on their way from Washington to New York during those 5 hours. Keep in mind that the Washington-to-New York trains leave on the hour and on the half-hour.

At the Washington station at 5:00 P.M., there is a train ready to leave for New York. Call it **A**:

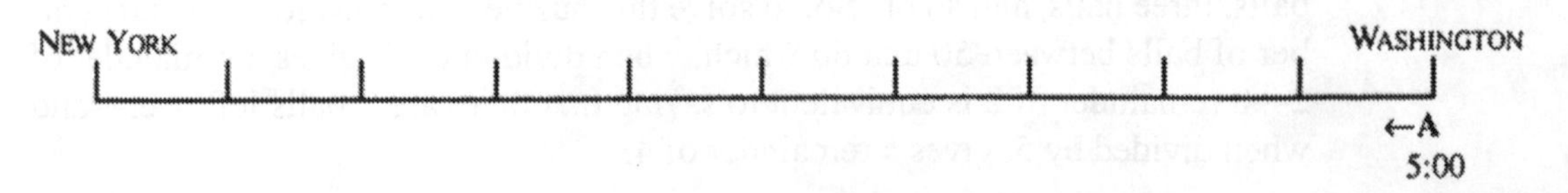

Obviously, the train that had left the Washington station a half hour earlier, at 4:30 P.M., will find itself a certain distance from the Washington station when **A** is about to leave. Call that train **B**:

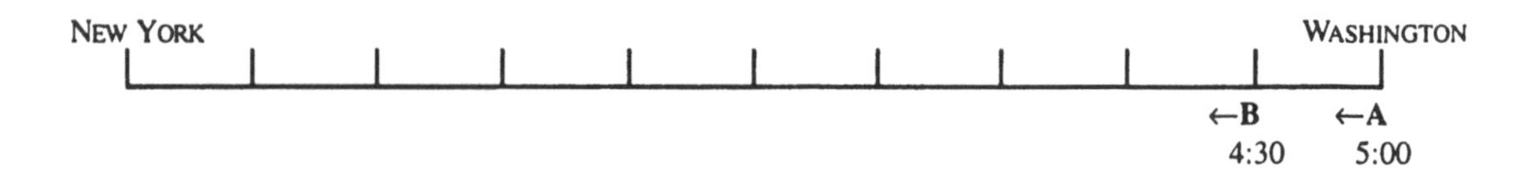

You can now complete the diagram showing the relative locations of all the trains from Washington bound for New York that left the Washington station between 12:00 P.M. and 5:00 P.M. as follows:

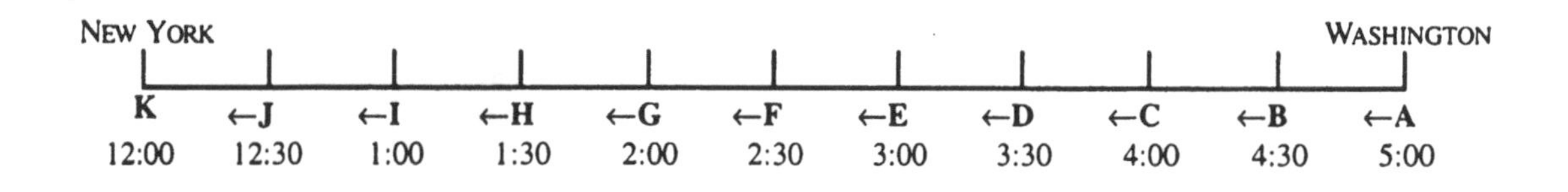

Now, when you left the New York station at 12:00 noon, you obviously missed passing the 12:00 o'clock **K**-train from Washington, because it was in the station when your train was leaving. But, as you can see from the diagram above, you passed all the others—the 12:30 **J**-train (i.e., the train that left Washington for New York at 12:30), the 1:00 **I**-train, the 1:30 **H**-train, the 2:00 **G**-train, the 2:30 **F**-train, the 3:00 **E**-train, the 3:30 **D**-train, the 4:00 **C**-train, the 4:30 **B**-train, and the 5:00 **A**-train. That makes 10 trains in all.

29 **Answer:** *One-third wine.*

Solution: The puzzle tells you that container **A** is half full of wine and that container **B**, which is twice the size of **A**, is one-quarter full of wine. First, draw the two containers, making **B** twice the size of **A**:

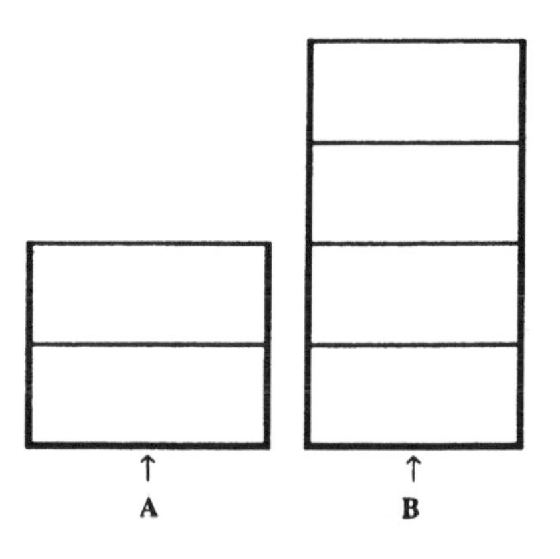

Now, when you fill half of **A** and one-quarter **B** with wine, the containers will look like this:

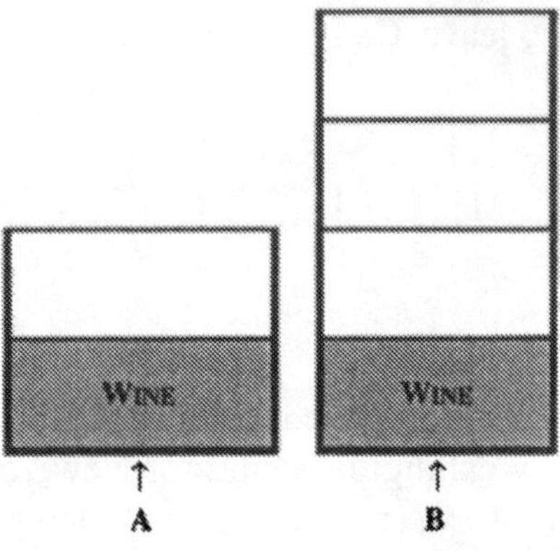

Notice that there is, in fact, the same amount of wine in the two containers. Now fill the containers with water:

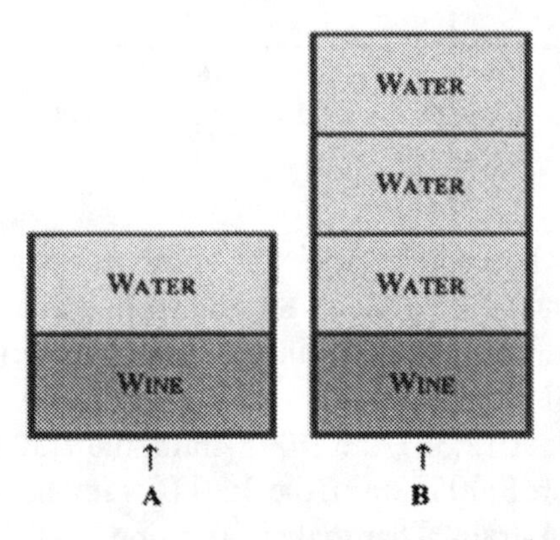

As can now be seen, **A** has two equal portions of wine and water, while **B** has three parts water and one part wine. Between the two containers, there are six equal parts in total—two parts wine and four parts water. Logically, a mixture of these two containers will contain two parts wine and four parts water. That is, in fact, what container **C** will have:

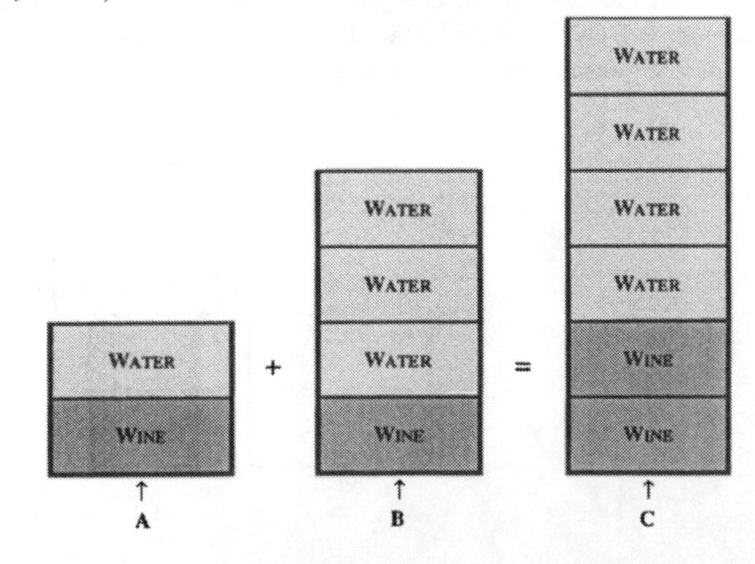

The wine and water in container **C** will, of course, be mixed up, not separated neatly like we have shown in the diagram above. But, in that mixture, wine will make up two parts out of its six, or $2/6$, and water will make up four parts out of its six, or $4/6$. In conclusion, C's mixture will have $2/6 = 1/3$ wine in it.

30 **Answer:** *3 mph.*

Solution: This classic puzzle seems to mislead many solvers, probably because they assume that the average speed is simply calculated by adding the hiker's two rates—2 mph and 6 mph—together (=8 mph) and then dividing by 2. They invariably come up with the incorrect answer of 4 mph.

As always, you can avoid pitfalls by reasoning on a step-by-step basis. Recall from your school years that:

$$\text{Distance} = \text{Rate} \times \text{Time}$$

OR (in abbreviated form)

$$D = R \times T$$

Now, calculate the time Danielle took to hike uphill. For the sake of convenience, assume that the distance uphill (and downhill, of course) is 1 mile. You can use any other distance; the reasoning and result will be the same. You are told that her rate uphill is 2 mph. So, her *time up* is:

$$D = R \times T$$

$$1 = 2 \times \textit{time up}$$

OR

$$\textit{time up} = 1/2$$

Thus, it would take Danielle $1/2$ hour to climb a 1-mile hill. Now, given that her rate downhill is 6 mph, calculate what her *time down* is:

$$D = R \times T$$

$$1 = 6 \times \textit{time down}$$

OR

$$\textit{time down} = 1/6$$

So, it would take Danielle $1/6$ hour to descend a 1-mile hill. Now, her total time for the entire trip is, of course, *time up* + *time down,* or $1/2 + 1/6 = 2/3$ hour. The total distance Danielle covered is 2 miles—1 mile up + 1 mile down. Therefore, her overall rate is:

$$D = R \times T$$

$$2 = R \times 2/3$$

$$R = 3 \text{ miles per hour}$$

As you can see, Danielle's overall, or average, rate of speed, which has been calculated by taking into account the overall distance she covered (=2 miles) and the overall time she took (=$2/3$ hour), is 3 mph, not 4 mph!

31 **Answer:** *7 dimes and 13 nickels.*

Solution: This puzzle could, of course, be solved by algebraic means (see Chapter 5). But it can also be reasoned out in terms of number logic. First, assume that Jim has an equal number of dimes and nickels in his pocket—10 of each. How much money does that make? Let's see.

➤ Scenario 1

10 dimes	=	$1.00
10 nickels	=	.50
Total	=	$1.50

The total is obviously too high, because Jim has only $1.35 in his pocket. So, clearly, fewer dimes are needed in the addition scenario. If a dime is taken away from the 10 dimes in scenario 1, then the number of nickels must be increased by one—because Jim has 20 coins in his pocket. This is what 9 dimes and 11 nickels add up to in money terms:

➤ Scenario 2

9 dimes	=	$ .90
11 nickels	=	.55
Total	=	$1.45

This total is still greater than $1.35. So, try reducing the number of dimes in Jim's pocket by two (from the 10 in scenario 1):

➤ Scenario 3

8 dimes	=	$ .80
12 nickels	=	.60
Total	=	$1.40

This total is still too high, but you are getting closer to the required total of $1.35. So, let's see what happens when the number of dimes in Jim's pocket is reduced by three (from the 10 in scenario 1):

➤ Scenario 4

7 dimes	=	$ .70
13 nickels	=	.65
Total	=	$1.35

As can be seen, scenario 4 contains the solution—Jim has 7 dimes and 13 nickels in his pocket.

32 **Answer:** *15 miles.*

Solution: This puzzle is an all-time favorite, which, like puzzle 30 above, leads many solvers astray. The first thing to do is figure out how much time it took the bike riders to cover the 20 miles between them. Both were traveling at the same constant rate of 10 mph toward each other. So, they would meet halfway: that is, after they had covered 10 miles of the 20 miles. Moving at 10 miles per hour, it therefore took 1 hour for each bike rider to cover the distance of 10 miles.

$$D = R \times T$$
$$10 = 10 \times T$$
$$T = 1$$

Now, consider how much distance the fly covered during that 1 hour. You are told that the fly went back and forth at 15 miles per hour during that 1 hour. So, that's it! In 1 hour the fly covered a total distance of 15 miles!

$$D = R \times T$$
$$D = 15 \times 1$$
$$D = 15$$

The reason why this puzzle gives many solvers difficulties is probably because the fly's distance is described in terms of a *back-and-forth* movement. The fly will cover a certain distance, not a certain direction, within the specified period of 1 hour! It doesn't matter if that distance can be mapped out as a straight line, or as a back-and-forth movement!

Incidentally, if you were fooled into believing that this puzzle was more complicated than the solution given above, then you are in really exceptional company! One of the greatest mathematicians of this century, the Hungarian-born professor of mathematics at Princeton University, John von Neumann (1903–1957), whose ideas led to the development of the modern computer, was once asked to solve this very puzzle at a cocktail party. The story purportedly goes like this. Von Neumann thought about the puzzle for a moment and then gave the correct answer. The person who had posed the puzzle to him was amazed, because he had always remarked that mathematicians constantly overlooked the simple way it can be solved, trying instead to solve it by a lengthy and complicated process using advanced mathematics (summing an infinite series). Von Neumann looked surprised and retorted, matter-of-factly, "Well, that's precisely how I solved it!"

33 **Answer:** *4 inches.*

Solution: This puzzle is another classic nut that people constantly find difficult to crack correctly. Needless to say, it would help enormously if you could actually see the books as they line up on a bookshelf. So, start with an appropriate diagram, identifying the three volumes in order from left to right as Volumes I, II, and III and putting them flush against each other. Indicate, as well, the widths of the covers—½ inch—and the widths of the pages in between two covers—2 inches—since we will need these to figure out how far the bookworm traveled:

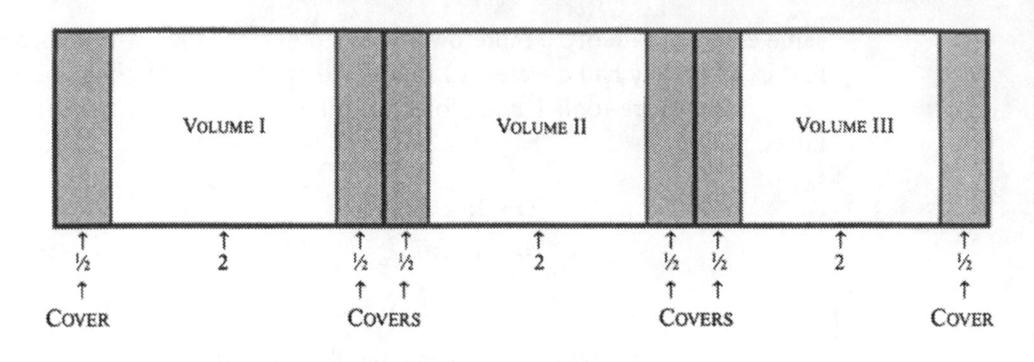

Next, label the relevant pages. You are told that the bookworm started on the first page of the first book to the left (=Volume I). Think carefully! If you were to pick up Volume I, where would the first page of that volume be? It would be on the right end of Volume I, as you look at Volume I on the bookshelf! If you do not quite see this, simply take three books, stack them up as shown, and then check out for yourself where the first page of the first book on the left is:

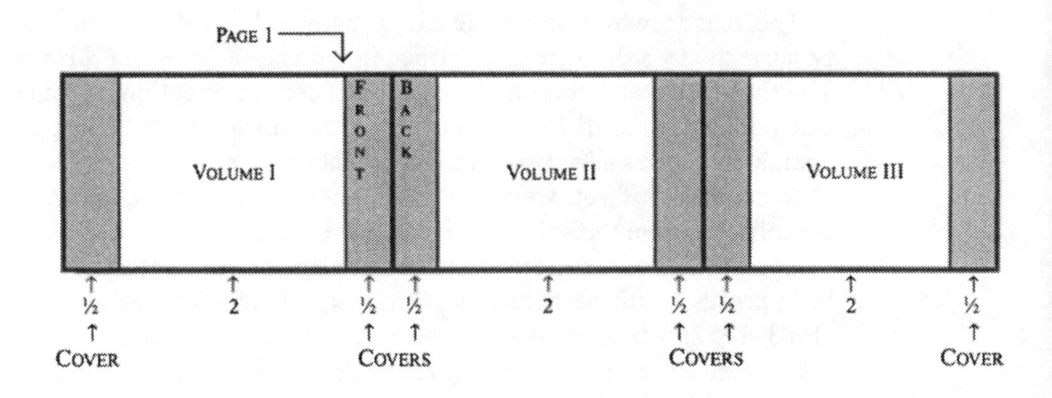

Then, you are told that the bookworm stopped at the last page of the last book to the right (=Volume III). Where is the last page of Volume III? It is on the left end of Volume III, as you look at that volume on the bookshelf! Once again, if you have difficulty seeing this, just stack three books up as shown, and then check out for yourself where the last page of the book on the right is:

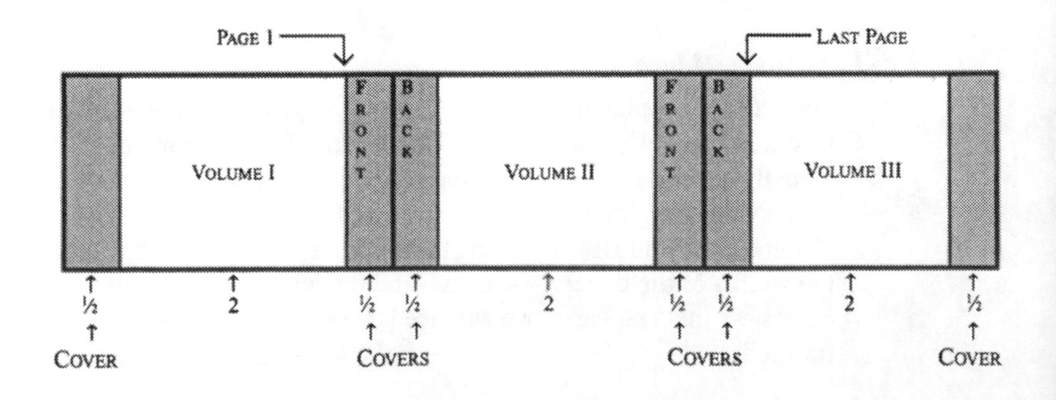

Now you can go ahead and calculate the distance the bookworm covered. The bookworm started on page 1 of Volume I, went through its front cover (=½ inch), then through the back cover of Volume II (=½ inch), then through the pages of Volume II (=2 inches), then through the front cover of Volume II (=½ inch), and finally through the back cover of Volume III (=½ inch), at which point it reached the last page of Volume III. Here's the bookworm's actual path:

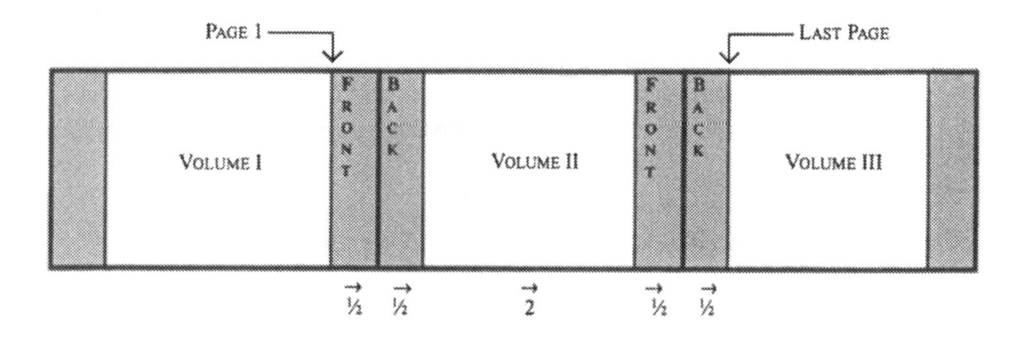

As you can see, the bookworm went through a distance of ½ + ½ + 2 + ½ + ½ = 4 inches. So, the bookworm traveled a total of 4 inches, from the front page of Volume I to the last page of Volume III.

34 **Answer:** *Option B.*

Solution: This is yet another classic puzzle that has been bewildering puzzle-solvers for eons! The best way to attack this puzzle is to do what you have been doing throughout this chapter: that is, use a concrete, step-by-step line of reasoning. The key to solving this puzzle is to actually count the salaries generated by the two options.

After the first year, with Option A you would receive just the $4,000. With Option B you would receive $2,000 after the first 6 months, but then you would get an increase of $200. So, during the last 6 months of that year, you would get $2,200 dollars. Adding the two half-years, you would get $4,200 at the end of the first year.

Year	Option A	Option B
1	$4,000	$4,200

Now, what income do both options generate after year 2? Well, with Option A you would get an increase of $800 for the year. So, you would end up earning $4,800. But with Option B, you would earn $2,400 in the first half-year—the $2,200 you would have started off with at the beginning of the year (=salary from the previous half-year) and the $200 raise you would have gotten for that half-year. Then, in the last 6 months you would get another increase of $200 on top of this new salary: that is, you would earn another $2,600 ($2,400 + $200). Adding the two half-years, you would get $5,000 at the end of the second year.

If you do not see this, consider Option B in its "sequential" components, 6 months at a time:

	Raise	Half-Year Salary	Total Salary at Year's End
Months 1–6	—	$2,000	—
Months 7–12	$200	$2,200	$4,200
Months 13–18	$200	$2,400	—
Months 19–24	$200	$2,600	$5,000

To summarize, after 2 years here's what the two options would generate:

Year	Option A	Option B
1	$4,000	$4,200
2	$4,800	$5,000

If you continue calculating the incomes generated by the two options in this way for, say, 4 more years, you would see that Option B actually generates more income, and is therefore the better option:

Year	Option A	Option B
1	$4,000	$4,200
2	$4,800	$5,000
3	$5,600	$5,800
4	$6,400	$6,600
5	$7,200	$7,400
6	$8,000	$8,200

5 Puzzles in Algebraic Logic

Algebra is not the frightening subject many people think it is. In a basic sense, it is no more than "arithmetic with letters," and like arithmetic, it has been the source of countless puzzles throughout the centuries. The main difference between solving puzzles in pure arithmetic and solving those in algebra lies in knowing how to set up an appropriate equation or series of equations.

The history of algebra began in ancient Egypt and Babylon. The Babylonians solved equations by essentially the same procedures taught today. This ancient knowledge found a home early in the Islamic world, where it was known as the "science of restoration and balancing." Indeed, the Arabic word for restoration, *al-jabr,* is the root of the word *algebra.* In the ninth century, the Arab mathematician Mohammed ibn-Musa al-Khowarizmi wrote one of the first algebra textbooks, a systematic exposé of the basic theory of equations, with both examples and proofs. By the end of the ninth century, the Egyptian mathematician Abu Kamil had formulated and proven the basic laws and identities of algebra.

This is actually a key chapter to master, since many of the puzzles in combinatory (Chapter 6), geometrical (Chapter 7), and time (Chapter 9) logic require that you know how to set up and solve equations. Moreover, the puzzles in code logic (Chapter 8) are algebraic in their essence, since they require that you find numbers corresponding to letters of the alphabet—the fundamental characteristic of algebraic logic.

How To...

The only background knowledge you will need to solve the puzzles in this chapter is how to set up and solve elementary equations with one unknown. There are, of course, many algebraic puzzles that can only be solved with the use of more than one unknown or variable. The intent of this chapter, though, is simply to help you understand the basic principles that must be grasped in order to be able to convert ordinary language into algebraic language. Once you have mastered these principles involving equations with one variable, you will be able to extend them to the solution of puzzles involving equations with two or more variables easily.

PUZZLE PROPERTIES

You will recognize an algebraic puzzle readily because it will invariably require that you find out the numerical value of a certain quantity, amount, measure, and so on (the length of a trout, the number of people in a group, someone's age, etc.), by telling you how this *unknown* value relates to other numbers and facts: for example, you might be told that someone's age will be twice what it is now in a specified number of years from the present, or that the total number of people in one group exceeds the number of people in another group by a certain amount.

Example 1 The following puzzle is typical of all algebraic puzzles.

> The other day my friend went fishing. She caught an enormous trout. It was 20 meters long plus half of its own length. How long was it?

The key to solving this, and all puzzles in algebraic logic, is to translate a verbal statement into its appropriate algebraic notation. So, the first thing to do is to let x stand for the *total length* of the trout. Why? Because that is what you are asked to find out. The letter x is called, appropriately enough, the *unknown*! The puzzle also tells you that the trout was 20 meters long plus *half of its own length.* The latter statement translates into $\frac{1}{2}x$. Do you see why?

If you've forgotten your algebra, a few concrete examples should convince you that this is the proper notation:

- If the trout were 30 meters long, then half of that length would be $\frac{1}{2} \times 30$.
- If the trout were 29 meters long, then half of that length would be $\frac{1}{2} \times 29$.
- If the trout were 28 meters long, then half of that length would be $\frac{1}{2} \times 28$.
- And so on.

Now, since the trout in question was actually x meters long, then half of that length would be $\frac{1}{2} \times x$, or more simply, $\frac{1}{2}x$.

Next, translate the relevant verbal statements into an equation. You are told that the trout's *total length* (x) was *20 meters* plus *half its own length* ($\frac{1}{2}x$). In algebraic language, this translates into $x = 20 + \frac{1}{2}x$:

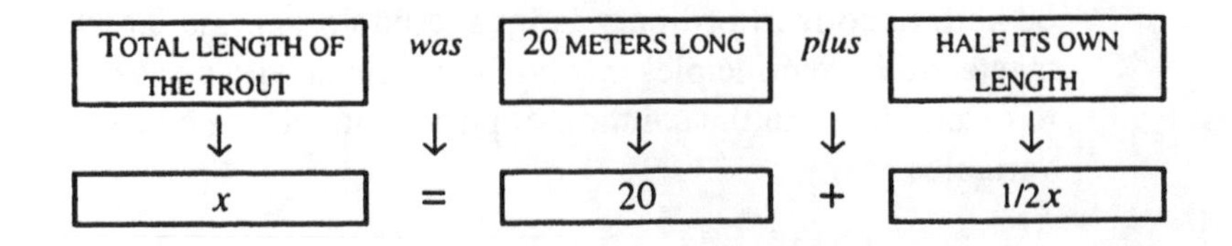
TOTAL LENGTH OF THE TROUT
was
20 METERS LONG
plus
HALF ITS OWN LENGTH
x
=
20
+
1/2x

Solving the equation, you'll get $x = 40$. This tells you that the trout was 40 meters long. Before leaving the puzzle, it is always wise to check your answer against the facts as stated in the puzzle. *One-half* of the *total length* of 40 is, of course, $\frac{1}{2} \times 40 = 20$. Adding this to the 20 meters as stated in the puzzle (*It was 20 meters long plus half of its own length*), $20 + 20 = 40$, yields the answer.

Example 2 Although the following puzzle is more complicated, it is solved in exactly the same way as the previous one, by translating each relevant statement into appropriate algebraic notation.

> The other day my brother asked me: "My dear sister, do you still have my chocolate bars?" Here's what I told him, to confuse him a little bit: "I gave half of them to mom, and a half a bar more to dad. Then I gave half of what was left, and a half a bar more to our dog. That left me with one chocolate bar, which I gladly ate myself." Can you figure out how many chocolate bars the brother originally gave his sister?

Once again, start by letting x stand for the quantity that you are asked to find out, namely the total number of chocolate bars that the brother originally gave to his sister. The sister tells you that she gave half of the original number of chocolate bars to her mom. How do you translate that statement into algebraic notation? As you discovered above in example 1, the algebraic equivalent of such a statement is $\frac{1}{2}x$. This is how many chocolate bars she gave to her mom.

The sister then said that, in addition to what she gave to her mom ($\frac{1}{2}x$), she gave *half a bar* to her dad. Algebraically, this means, of course, that $\frac{1}{2}$ bar is to be added to $\frac{1}{2}x$ bars. So, in total, she gave away $\frac{1}{2}x + \frac{1}{2}$ bars to her mom and dad:

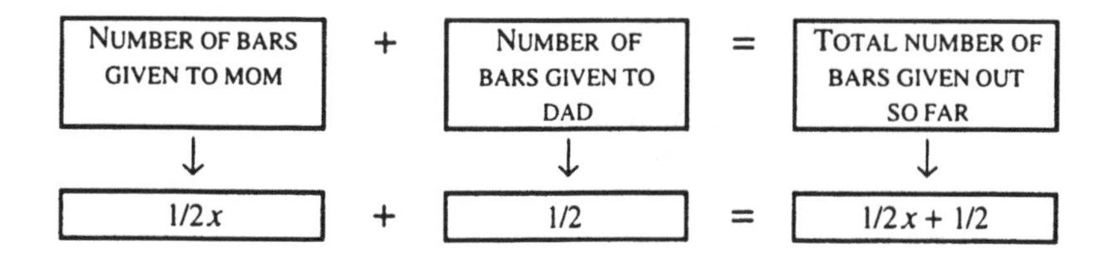

Her next statement concerns the number of bars she gave to her dog out of *what was left*. So, first you will have to figure out how many bars were left from the original number of bars, x, after she had given the $\frac{1}{2}x + \frac{1}{2}$ bars to her mom and dad. This means, of course, that you will have to take away the amount given to her mom and dad, $\frac{1}{2}x + \frac{1}{2}$, from the original total of x. Do you see this? For the sake of concreteness, suppose that the original number of bars was 10; then she would have given half of that, or 5, and $\frac{1}{2}$ more, or $5\frac{1}{2}$ in total, to her mom and dad. To find out how many bars would have been left over from the original 10, simply subtract the 5½ bars from the original 10 bars: that is, $10 - 5\frac{1}{2} = 4\frac{1}{2}$. So, there would have been 4½ bars left over for the sister to give to her dog.

Now, back to our puzzle. In algebraic notation *"what was left"* translates into:

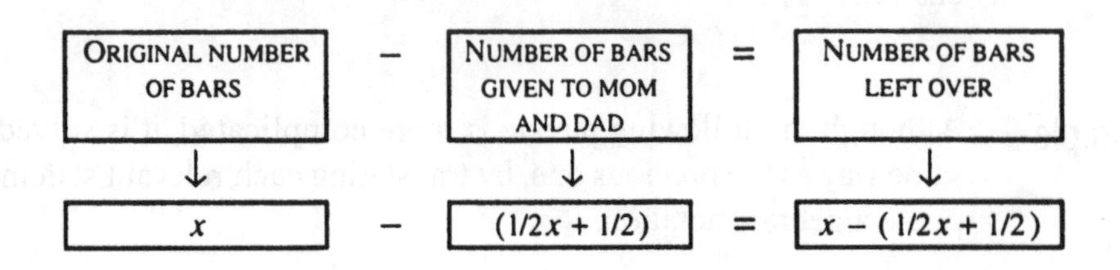

At this point, it will facilitate matters significantly if you *reduce* the algebraic expression to the right: $x - (\frac{1}{2}x + \frac{1}{2})$. From your school algebra, you should be able figure out that this expression reduces to $\frac{1}{2}x - \frac{1}{2}$. This now makes it much easier to continue.

The sister gave half of this number—$\frac{1}{2}(\frac{1}{2}x - \frac{1}{2}) = \frac{1}{4}x - \frac{1}{4}$ (in reduced form)—and ½ bar more—$(\frac{1}{4}x - \frac{1}{4}) + \frac{1}{2} = \frac{1}{4}x + \frac{1}{4}$ bars (in reduced form)—to her dog. To keep track of what she has given out so far, add the number of bars she gave to her mom and dad to the number she gave to her dog:

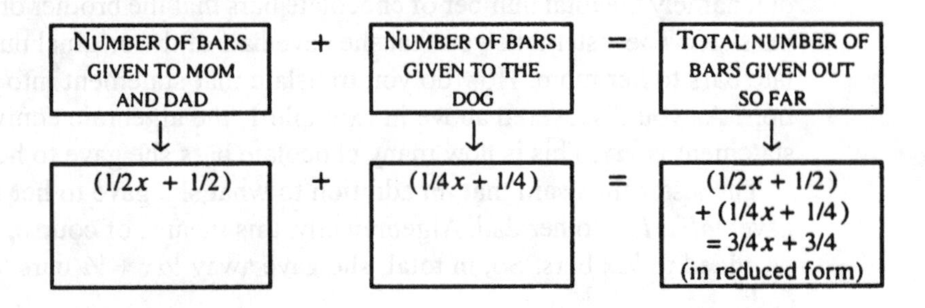

The last statement that is relevant to setting up an appropriate equation is that after the sister had given the above total number of bars out—$\frac{3}{4}x + \frac{3}{4}$—to her mom, dad, and the dog, she was still left with 1 chocolate bar. Calculate in algebraic terms how many bars were in fact left over after she had given out the $\frac{3}{4}x + \frac{3}{4}$ chocolate bars to mom, dad, and the dog. As before, this means subtracting $\frac{3}{4}x + \frac{3}{4}$ bars from the original number of x bars:

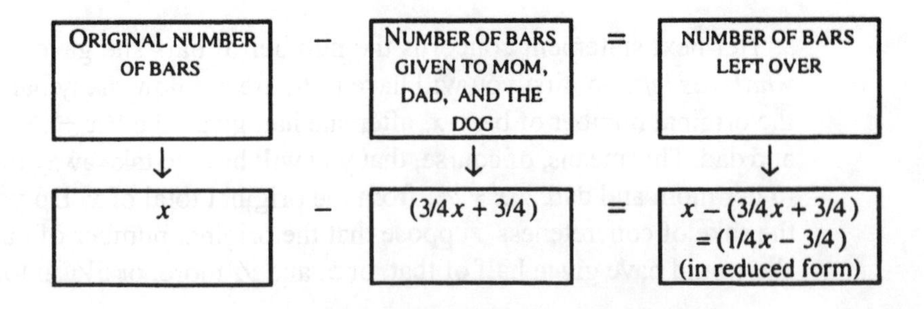

It is the expression to the right that is, of course, equal to 1: that is, $\frac{1}{4}x - \frac{3}{4} = 1$. Do you see this? The expression $\frac{1}{4}x - \frac{3}{4}$ is the number of bars left over when you take away the number the sister has given out to her mom, dad, and the dog from the total number of bars. The sister tells you exactly what this quantity is equal to—namely, to the *one* chocolate bar that she ate. Solving for x, you'll get $x = 7$. This tells you that the brother originally gave his sister seven chocolate bars.

As before, check your solution against the relevant statements in the puzzle. The sister gave half of the bars to her mom: that is, she gave her mom half of 7, which is $3\frac{1}{2}$ bars. Then, she gave $\frac{1}{2}$ bar to her dad. Altogether, she gave $3\frac{1}{2} + \frac{1}{2} = 4$ bars to her mom and dad. Then, she gave half of what was left over, $7 - 4 = 3$, or $\frac{1}{2} \times 3 = 1\frac{1}{2}$ bars, and $\frac{1}{2}$ bar more, or $1\frac{1}{2} + \frac{1}{2} = 2$ bars, to the dog. So, she gave her mom and dad 4 bars and the dog 2 bars. Altogether, she gave her mom, dad, and the dog $4 + 2 = 6$ bars. The bar she ate herself makes 7 bars in total.

Summary

Solving puzzles in algebraic logic is essentially an exercise in translation. The basic strategy is to translate each relevant verbal statement into its algebraic counterpart, as illustrated in examples 1 and 2. So, when attacking puzzles in algebra:

- Start off by letting x (or some other letter of the alphabet) stand for whatever it is you're looking for or trying to determine.
- Take each relevant verbal statement and translate it into its corresponding algebraic notation.
- Set up the equation that is implied in one of the statements of the puzzle, always looking to make any changes along the way (reducing, simplifying, etc.) that render the mechanical task of actually setting it up easier.
- Solve for x.
- Check out your answer against the statements in the puzzle, in order to verify its accuracy.

Puzzles 35–44

Answers, along with step-by-step solutions, can be found at the end of the chapter.

35 Nora, who works for a construction company, put a brick this morning on one pan of a weighing scale. She got a balance when she put three-quarters of another brick of the same kind and three-quarters of a kilogram weight on the other pan. How much did the brick weigh?

36 A pencil and eraser together cost 55¢. The pencil costs 50¢ more than the eraser. How much does the eraser cost?

37 Mary had a pocketful of dollar bills yesterday, so she went on a spending spree. At an office supply store, she spent half of the money on a new plastic briefcase and, being of a generous nature, gave a dollar to a beggar outside the store. Then, she had lunch at a trendy new restaurant, spending half of the remaining dollars and tipping the waiter 2 dollars. Finally, she stopped at a bookstore to buy some books. There, she spent half her remaining dollars, as well as putting 3 dollars in a charity box inside the store. This left her with barely one dollar in her pocket. How much money did she start out with?

38 Lucia is a wonderful grandmother. Everybody loves her. Her age is between 50 and 70. Each of her sons has as many sons as brothers. The combined number of Lucia's sons and grandsons equals her age. How old is Lucia?

39 The sum of the ages of Jack and Jill is 60 years. Jack is 3 years older than half Jill's age. How old is each one?

40 Jimmy has a total of $1.08 in his pocket in pennies, nickels, and dimes. He has twice as many dimes as pennies, and three times as many nickels as pennies. How many pennies, nickels, and dimes does Jimmy have in his pocket?

41 Tracy and Josephine are sisters. Their ages add up to 11 years, although Tracy is 10 years older than her sister. How old are the two?

42 Two automobiles are 425 miles apart. They start driving toward each other at the same time. The first one drives constantly at 45 miles per hour, the second at 40 miles per hour. How far will each car have traveled when they meet?

43 Millicent wants to buy her daughter a chain-link toy to help her develop her spatial reasoning abilities. The idea of the toy is to link up balls of the same color. It has 10 silver balls, 2 blue balls, 1 white ball, and a number of red and turquoise balls. How many balls does the chain-link toy have altogether, if the number of red balls equals one-third the entire chain minus three balls, and the number of turquoise balls is one-half the number of red balls plus twice the number of silver balls?

44 Mark bought some candy last week at 92¢ per pound. If the price had been 4¢ per pound lower, he could have bought one pound more for the same money. How much did he spend for the candy?

Answers and Solutions

35 **Answer:** *3 kilograms.*

Solution: This puzzle is similar to example 1 above. The first thing to do is let x stand for the total weight of the brick that Nora put on one pan—say the left pan. Nora then put a brick that was *three-quarters of that weight* on the right pan. The weight of that brick was, therefore, $\frac{3}{4}x$. Finally, she also put *three-quarters* of a kilogram weight on the right pan. Thus, the total weight that she put on the right pan was $\frac{3}{4}x + \frac{3}{4}$.

At that point Nora got a balance. So, the weight of the brick on the left pan (x) equaled the total weight of the items Nora put on the right pan ($\frac{3}{4}x + \frac{3}{4}$):

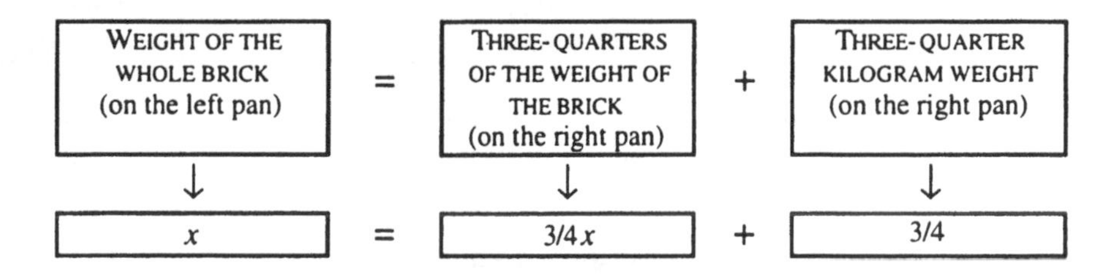

Solving the equation $x = \frac{3}{4}x + \frac{3}{4}$, you'll get: $x = 3$. So, Nora's brick weighed 3 kilograms.

Check this. Three-quarters of 3 kilograms is $2\frac{1}{4}$ kilograms. This is the weight of the brick that was put on the right pan. Adding a $\frac{3}{4}$ kilogram weight to the right pan makes the total weight on that pan $2\frac{1}{4}$ kilograms + $\frac{3}{4}$ kilograms = 3 kilograms.

36 **Answer:** *The eraser costs $2\frac{1}{2}$¢.*

Solution: Start by letting one of the two items be represented by x. The most logical choice is the eraser, because you are told that the pencil costs 50¢ *more* than what the eraser costs. So, if the eraser costs x cents, then the pencil costs 50¢ more than that, or $x + 50$. If you do not see this, consider a few concrete examples: if the eraser costs 1¢, then the pencil will cost 1¢ + 50¢ *more* (=51¢); if the eraser costs 2¢, then the pencil will cost 2¢ + 50¢ *more* (=52¢); and so on.

You are told that the price of the eraser and the pencil together is 55¢:

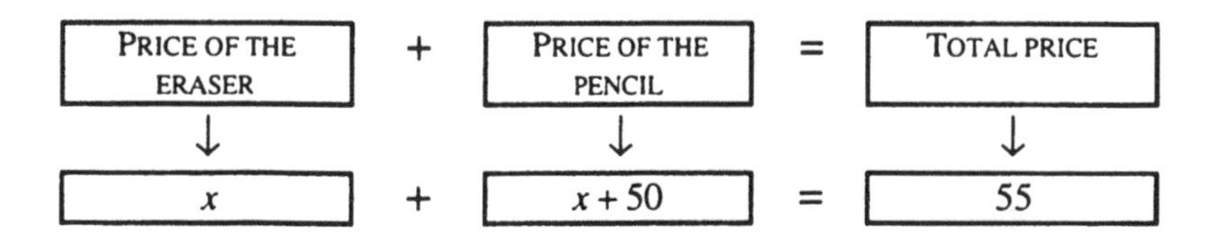

Solving the equation $x + x + 50 = 55$, or $2x + 50 = 55$, you'll get: $x = 2\frac{1}{2}$. So, the eraser costs $2\frac{1}{2}$¢ and the pencil $52\frac{1}{2}$¢ (=$2\frac{1}{2} + 50$).

Let's confirm this by simply adding the price of the eraser to the price of the pencil: $2\frac{1}{2}$¢ + $52\frac{1}{2}$¢ = 55¢.

37 **Answer:** *\$42.*

Solution: This puzzle is a version of example 2 above. The first thing to do is to let x stand for the total number of dollar bills Mary had in her pocket yesterday before she started out on her spending spree.

At the office supply store, Mary spent the following amount:

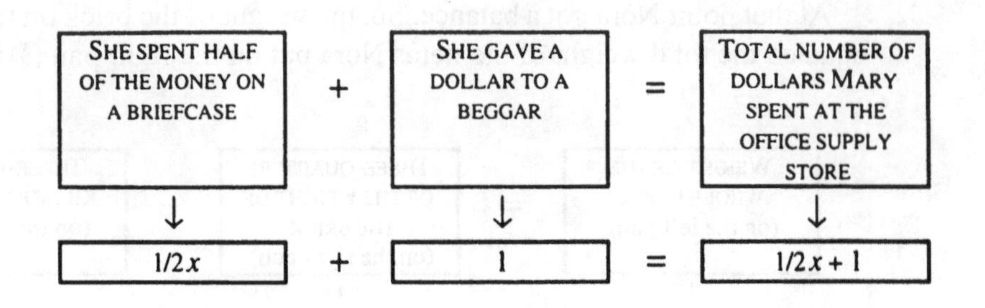
SHE SPENT HALF OF THE MONEY ON A BRIEFCASE
+
SHE GAVE A DOLLAR TO A BEGGAR
=
TOTAL NUMBER OF DOLLARS MARY SPENT AT THE OFFICE SUPPLY STORE
1/2x
+
1
=
1/2x + 1

How many dollars did she have in her pocket after she went away? To find out, simply subtract what she spent at the office supply store from the money she started out with: $x - (\frac{1}{2}x + 1) = \frac{1}{2}x - 1$. This is how many dollars *remained* in her pocket when she left for the restaurant.

At the restaurant, Mary then spent the following amount:

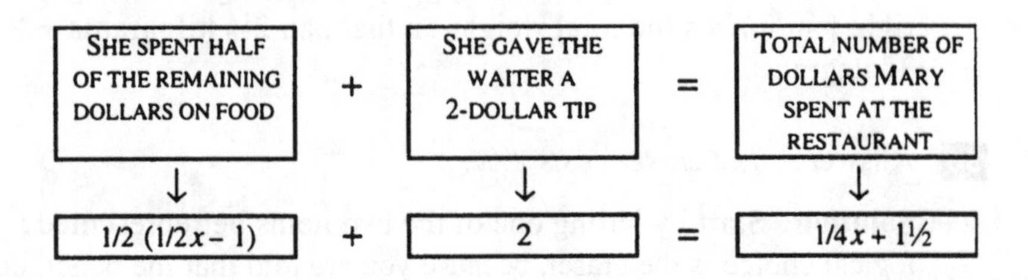
SHE SPENT HALF OF THE REMAINING DOLLARS ON FOOD
+
SHE GAVE THE WAITER A 2-DOLLAR TIP
=
TOTAL NUMBER OF DOLLARS MARY SPENT AT THE RESTAURANT
1/2 (1/2x – 1)
+
2
=
1/4x + 1½

If you are having difficulties seeing the reason behind the algebraic expression $\frac{1}{2}(\frac{1}{2}x - 1)$, just think of what each part stands for:

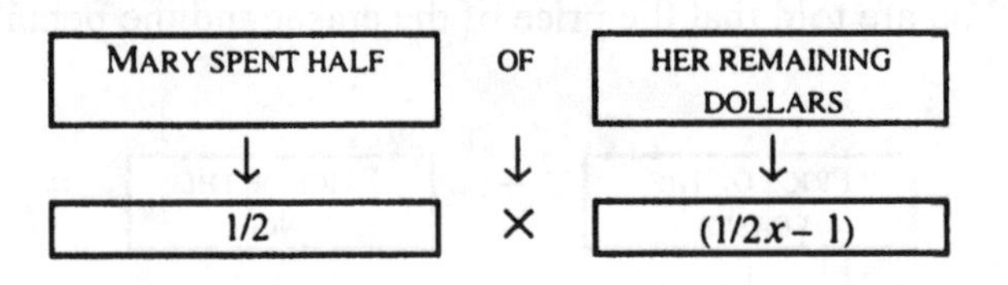
MARY SPENT HALF
OF
HER REMAINING DOLLARS
1/2
×
(1/2x – 1)

Now, before determining how much Mary spent at the bookstore, you will have to figure out how much money she had in her pocket after she left the restaurant. To do so, all you have to do is subtract what she spent at the office supply store and at the restaurant from the number of dollars she started out with. First, add up what she spent at the store and at the restaurant:

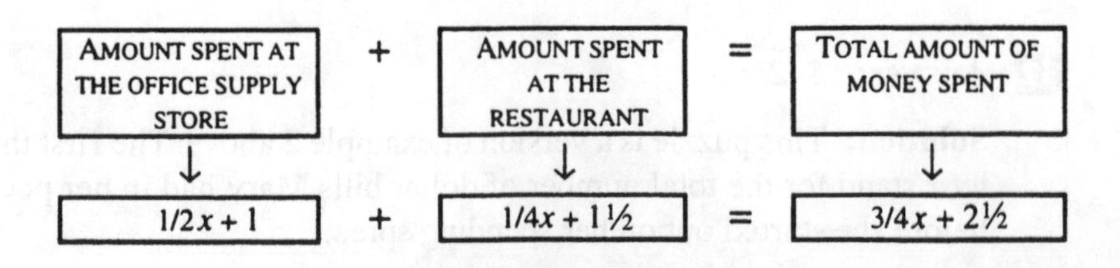
AMOUNT SPENT AT THE OFFICE SUPPLY STORE
+
AMOUNT SPENT AT THE RESTAURANT
=
TOTAL AMOUNT OF MONEY SPENT
1/2x + 1
+
1/4x + 1½
=
3/4x + 2½

Then, subtract this number from what she started out with: $x - (\frac{3}{4}x + 2\frac{1}{2}) = \frac{1}{4}x - 2\frac{1}{2}$. This is how much money Mary had in her pocket when she left for the bookstore.

At the bookstore, Mary spent the following amount:

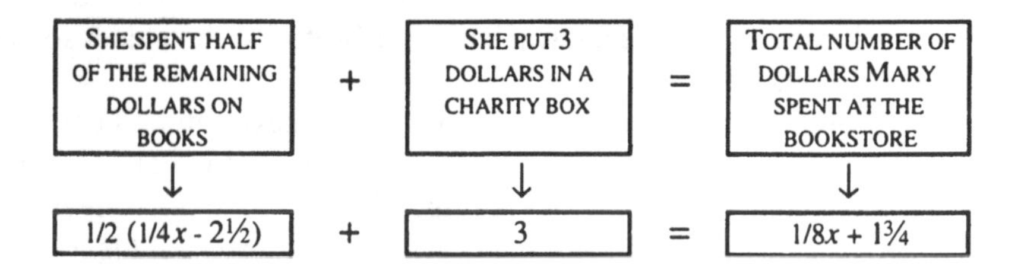

So, how much did Mary spend in total at the three places? Let's find out:

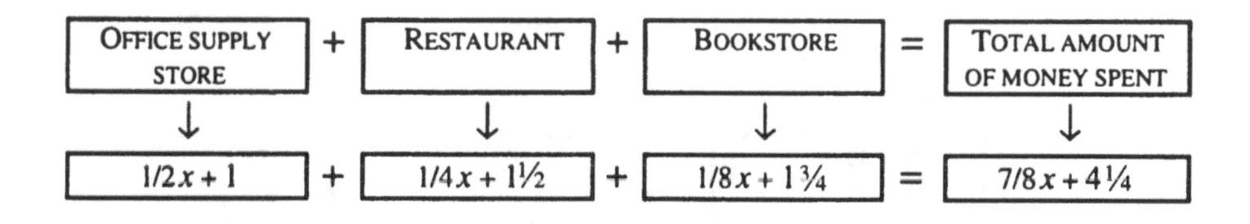

How much money remained in her pocket after her spending spree? To find out, subtract the total amount that Mary spent from the amount she started out with: $x - (\frac{7}{8}x + 4\frac{1}{4}) = \frac{1}{8}x - 4\frac{1}{4}$. The puzzle tells you that this amount is equal to 1 dollar: $\frac{1}{8}x - 4\frac{1}{4} = 1$. Solving this equation, you'll get: $x = 42$. So, Mary had $42 in her pocket at the beginning of yesterday.

See if this checks out. She spent half of the $42, or $21, on a new plastic briefcase. She then gave a dollar to a beggar. So, she spent a total of $22 at the office supply store. This left her with $42 – $22, or $20, in her pocket. At the restaurant, she spent half of the $20, or $10, on food. She tipped the waiter $2. In total, she spent $12 there. What did this leave her with? Well, remember that she spent $22 at the office supply store and $12 at the restaurant. This adds up to $34, leaving her with $42 – $34, or $8, in her pocket. Finally, at the bookstore she spent half of her remaining $8, or $4, and put $3 dollars in a charity box in the store. In total, she spent $7 there. So, altogether, Mary spent: $22 (at the office supply store) + $12 (at the restaurant) + $7 (at the bookstore) = $41. Adding the dollar that remained in her pocket to the $41 she spent equals $42 in total.

38 **Answer:** *Lucia is 64 years old.*

Solution: This is a really tricky puzzle. If you let x represent Lucia's age, then you probably got nowhere. But by now you should know that a good puzzle solver never gives up! So, try a different approach. Let x stand instead for the number of sons Lucia has. Let's see where this leads.

If x is the number of sons she has, then the number of brothers each son has is $x - 1$. If you do not see this, suppose that Lucia has 5 sons—Andy, Bill,

Charley, Dick, and Frank. How many brothers does Andy have? He has four brothers—Bill, Charley, Dick, and Frank. How many brothers does Bill have? He has four brothers—Andy, Charley, Dick, and Frank. And so on. Clearly, the number of brothers each of Lucia's five sons has equals one less (himself) than the total number of her sons, or $5 - 1$.

You are also told that each of Lucia's sons has, himself, as many sons as he has brothers. Since each son has $x - 1$ brothers, then each son also has $x - 1$ sons of his own.

Now, how many grandsons does Lucia have altogether? Well, if she has x sons, and each one has $x - 1$ sons of his own, then altogether she will have $x(x - 1)$ grandsons. Do you see this? If not, reconsider the concrete example above. If Lucia has 5 sons, then each one has 4 sons of his own. So, altogether Lucia has $5 \times 4 = 20$ grandsons. In general, if there are x sons in a family, and each one has $x - 1$ sons of his own, then altogether there will be $x(x - 1)$ grandsons in that family.

Finally, you are told that the combined number of Lucia's sons and grandsons equals her age, which is a number between 50 and 70:

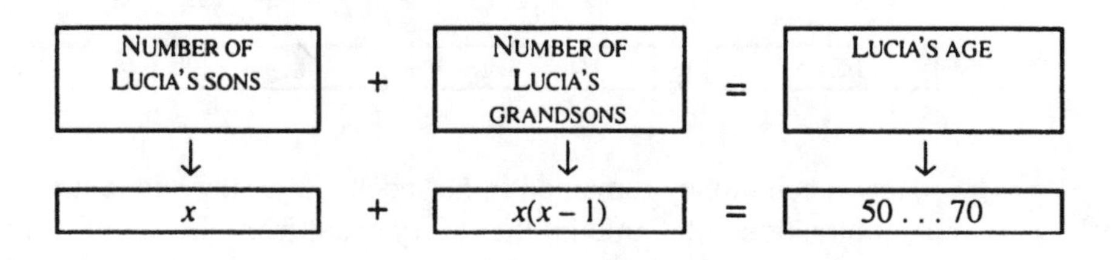

Simplifying the equation $x + x(x - 1) = 50 \ldots 70$, you'll get $x^2 = 50 \ldots 70$. The number we are looking for, therefore, is a square number (x^2) between 50 and 70. The only square number between 50 and 60 is 64: $8^2 = 64$. So, Lucia is 64 years old.

Check this answer out. Lucia has 8 sons (if $x^2 = 64$, then $x = 8$). Each son has 7 sons ($x - 1 = 7$). Altogether Lucia has $8 \times 7 = 56$ grandsons. If you add the number of sons and grandsons together—8 + 56—you'll get Lucia's age of 64.

39 **Answer:** *Jack is 22 years old and Jill is 38 years old.*

Solution: It doesn't matter whose age, Jack's or Jill's, you represent with x. Select Jack as the one who is x years old. Then Jill is $60 - x$ years of age. Do you see this? Suppose, for instance, that Jack's age is 20. You are told that the sum of the two ages equals 60. Therefore, Jill's age is 40—which is, of course, 60 – 20. If Jack's age were 35, then Jill's age would be 25—which is 60 – 35. And so on. Jack's age is x; so Jill's age is $60 - x$.

Now, consider the second statement in the puzzle: *Jack is 3 years older than half Jill's age.* What is *half Jill's age*? Jill's age is $60 - x$; so, half of that is $\frac{1}{2}(60 - x)$. Three years older than that can be represented simply by adding 3 years to the expression: that is, $\frac{1}{2}(60 - x) + 3$. This is what Jack's age equals:

$$x = \frac{1}{2}(60 - x) + 3$$

Solving for x, you'll get: $x = 22$. So, Jack is 22 years old and Jill is 38 years old ($60 - x = 60 - 22$). Check this. The sum of their two ages is, indeed, $22 + 38 = 60$ years. Half Jill's age is ½ of 38 = 19. And Jack, at 22, is indeed 3 years older than that.

40 **Answer:** *Jimmy has 3 pennies, 9 nickels, and 6 dimes.*

Solution: Start by letting x stand for the number of pennies Jimmy has in his pocket. Now, translate each verbal statement into its corresponding algebraic form:

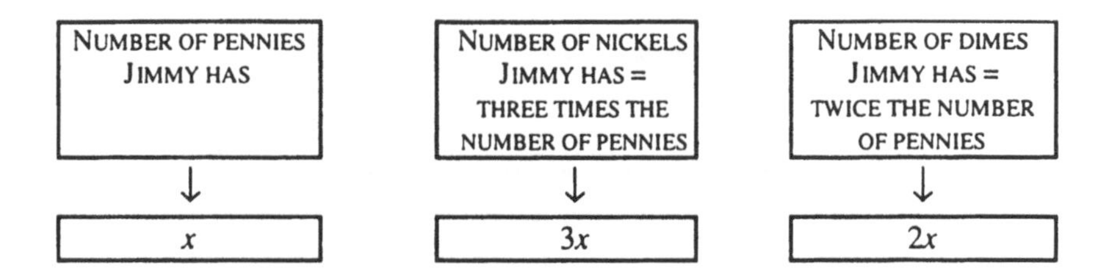

What is the value of a penny, a nickel, and a dime? A penny is worth \$.01, a nickel \$.05, and a dime \$.10. So, x pennies are worth \$.01x; $3x$ nickels \$.05($3x$); and $2x$ dimes \$.10($2x$). The puzzle tells you that these add up to \$1.08: $.01x + .10(2x) + .05(3x) = 1.08$.

Solving for x, you'll get: $x = 3$. So, Jimmy has 3 pennies, 9 nickels, and 6 dimes. Now, see if this works out: 3 pennies are worth \$.03; 9 nickels are worth $9 \times .05 =$ \$.45; and 6 dimes are worth $6 \times .10 =$ \$.60. Adding these values up, you'll get: \$.03 + \$.45 + \$.60 = \$1.08.

41 **Answer:** *Josephine is ½ year old, and Tracy is 10½ years old.*

Solution: This puzzle is similar to previous puzzles. So, it requires little commentary here. Start by letting x stand for Josephine's age. If her age is x years, then Tracy's age is 10 years more than that, or $x + 10$.

Now, you are told that the sum of their ages is 11 years:

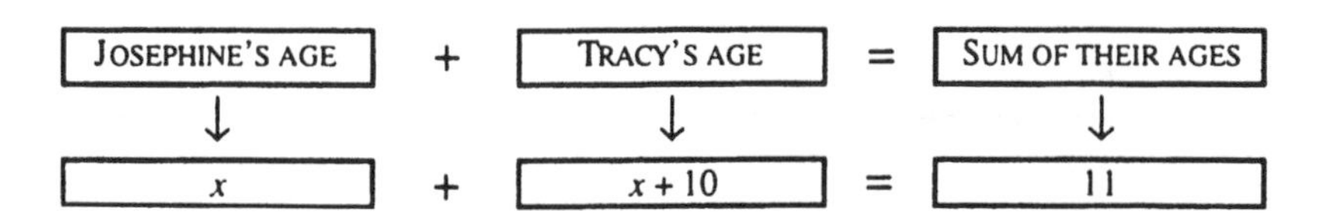

Solving the equation $x + x + 10 = 11$ (or $2x + 10 = 11$), you'll get $x = ½$. So, Josephine is ½ year old, and Tracy is 10½ ($= x + 10 = ½ + 10$) years old.

Adding Josephine's age of ½ to Tracy's age of 10½ adds up to 11—thus confirming the solution.

42 **Answer:** *One car will have traveled 225 miles and the other one 200 miles when they meet.*

Solution: Start by drawing a visual representation of what the puzzle says: that is, draw a line with two points on it, **A** and **B**, 425 miles apart:

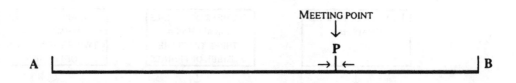

One car starts driving from **A** and the other from **B** at the same time. At a certain point—call it **P**—the two cars will meet:

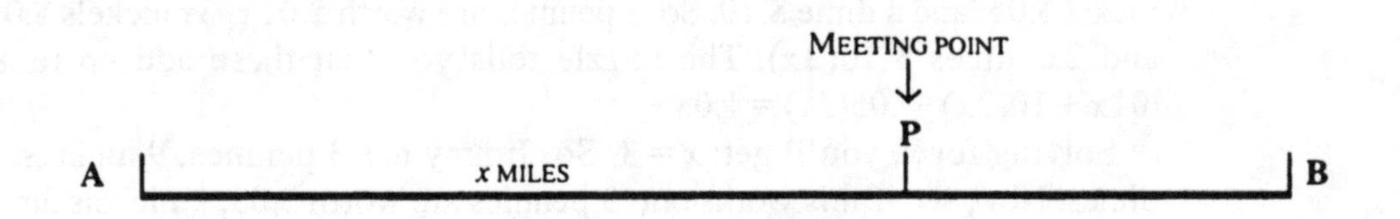

Now, represent the distance traveled by the car coming from **A** to **P** as x miles:

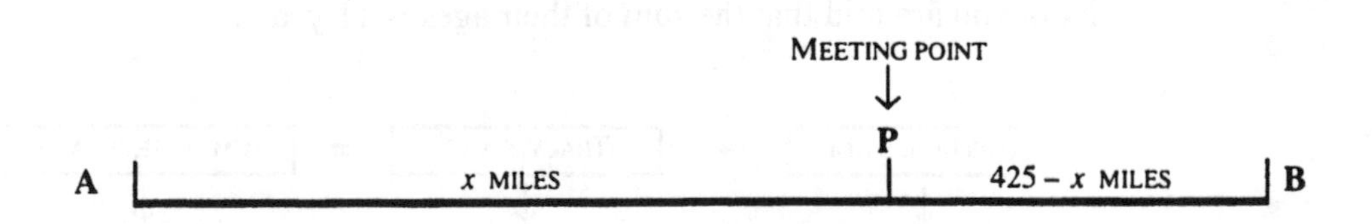

What distance will the other car have traveled from **B** to point **P** in the same amount of time? You know that there are 425 miles between **A** and **B**. And you have marked off the distance from **A** to **P** as x miles. So, the remainder of the distance—from **P** to **B**—is whatever is left over from 425: namely, $425 - x$:

MEETING POINT

P

A x MILES 425 – x MILES B

If you have difficulty seeing this, think of a few concrete examples. Assume, for instance, that the distance covered by the car from **A** to the meeting point **P** is 300 miles. Then, clearly, the other car will have traveled $425 - 300 = 125$ miles from **B** to that meeting point. If the distance covered by the car from **A** to **P** is 350 miles, then the other car will have traveled $425 - 350 = 75$ miles from **B** to **P**. And so on.

Distance equals rate multiplied by time:

$$D = R \times T$$

You are told what the rate of speed of each car is:

CAR FROM **A**:
$R = 45$ mph

CAR FROM **B**:
$R = 40$ mph

The car from **A** will have covered a distance of x miles to point **P**, and the car from **B** a distance of $425 - x$ miles. Therefore, the time it will have taken the car from **A** to reach **P** is:

CAR FROM **A**:

$$D = R \times T$$
$$x = 45 \times T$$
$$x/45 = T$$

And the time it will have taken the car from **B** to reach **P** is:

CAR FROM **B**:

$$D = R \times T$$
$$425 - x = 40 \times T$$
$$(425 - x)/40 = T$$

The two cars will take an equal amount of time to reach point **P**—one might have traveled further than the other, but it will have taken it the same amount of time to get there. Therefore, the two equations above for T are equal to each other:

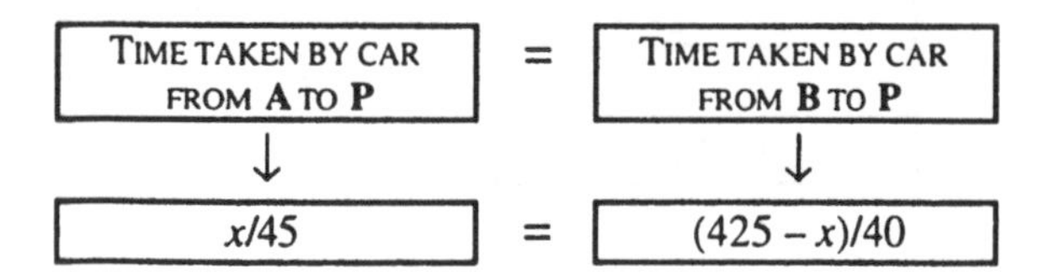

Solving for x, you'll get: $x = 225$. So, the car from **A** will have traveled 225 miles and the one from **B** 200 miles ($=425 - x = 425 - 225 = 200$) when they meet at point **P**.

You can check this out as follows:

CAR FROM **A**:

$$D = R \times T$$
$$225 = 45 \times T$$
$$5 = T$$

This tells you that it will have taken the car from **A** 5 hours to get to the meeting point **P**. This is, of course, the amount of time that the other car would also have taken to reach point **P** from **B**. And, in fact, this checks out:

CAR FROM **B**:

$$D = R \times T$$
$$200 = 40 \times T$$
$$5 = T$$

43 **Answer:** *57 balls.*

Solution: Start by letting x stand for the total number of balls in the toy. You are told that there are 10 silver balls, 2 blue balls, and 1 white ball on the chain. Altogether, this makes 13 balls. Then, you are told that the number of red balls on the chain is $\frac{1}{3}$ of the entire chain, minus three balls. This means that if there are x balls on the chain, then there are, obviously, $\frac{1}{3}x - 3$ red balls on it:

NUMBER OF RED BALLS

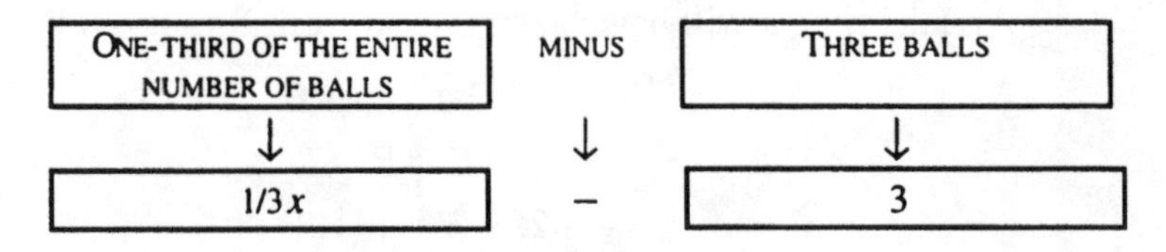

You are also told that the number of turquoise balls on the chain is $\frac{1}{2}$ the number of red balls ($\frac{1}{3}x - 2$) plus twice the number of silver balls ($2 \times 10 = 20$): that is, $\frac{1}{2}(\frac{1}{3}x - 3) + 20$:

NUMBER OF TURQUOISE BALLS

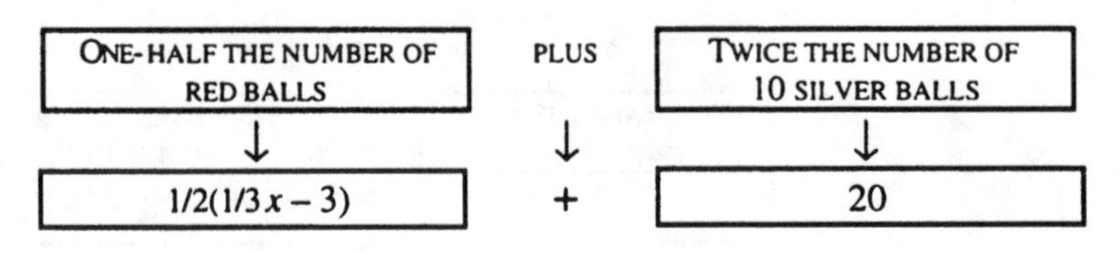

Altogether, the silver, blue, white, red, and turquoise balls add up to x—the total number of balls:

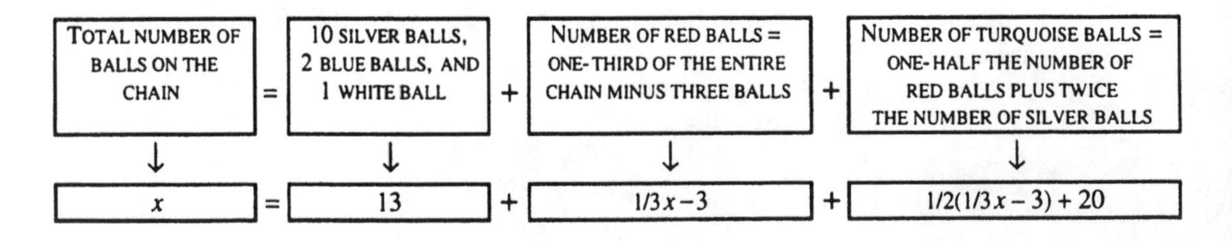

This equation simplifies to $3x = 171$. Solving for x, you'll get: $x = 57$ balls. Check this out. There are 10 silver balls, 2 blue balls, 1 white ball, 16 red balls ($\frac{1}{3}x - 3 = 19 - 3 = 16$), and 28 turquoise balls ($\frac{1}{2}[\frac{1}{3}x - 3] + 20 = 8 + 20 = 28$). Altogether, this makes: $10 + 2 + 1 + 16 + 28 = 57$.

44 **Answer:** *$20.24.*

Solution: Start by letting x represent the number of pounds of candy that Mark bought at 92¢ per pound. So, he spent $\$.92x$ for the candy.

Now, the puzzle tells you that if the price had been 4¢ per pound less than what the candy actually cost—that is, 88¢ per pound—then Mark could have bought one more pound of candy—$x + 1$ pounds—for the same money. The cost of $x + 1$ pounds at 88¢ per pound would have been, of course, $\$.88(x + 1)$:

$$.92x = .88(x + 1)$$

Solving for x, you'll get: $x = 22$. So, Mark bought 22 pounds of candy. At 92¢ per pound, he spent: $.92 \times 22 = \$20.24$.

Let's see if this checks out. At 4¢ per pound less—that is, at 88¢ per pound—Mark would have been able to buy 23 pounds of candy—that is, one pound more. The cost would, however, have been the same: $.88 \times 23 = \$20.24$.

6 Puzzles in Combinatory Logic

The study of how things can be arranged falls under the branch of mathematics known as *combinatorial analysis,* or *combinatorics.* Throughout history, puzzles in combinatory logic have provided mathematicians with pivotal insights into the laws of chance, the properties of numbers, and the nature of sets. This is why the great German philosopher, mathematician, and statesman Gottfried Leibniz (1646–1716) characterized the overall art of logical reasoning as an *ars combinatoria,* a "combinatory art." Today, combinatorics has important applications to the design and operation of computers as well as to the physical and social sciences. Indeed, in any area where the possible arrangements of a finite number of elements play a role, combinatorial analysis is useful.

The puzzles in this chapter involve the arrangement of numbers, colors, weights, and so on. In other puzzle books, you will find that some of these puzzles have been classified under different rubrics. The decision to include them here has been motivated simply by the fact that they are based on some combinatory pattern or principle. Not included in this chapter are puzzles involving boards (chess boards, domino boards, etc.) or puzzles dealing with tiling and playing card arrangements. There exist, in actual fact, numerous kinds of puzzles based on general principles of combinatorics. The reasoning processes, and the broad lines of attack illustrated in this chapter, are applicable to a large number of cases.

How To...

Solving puzzles in combinatory logic requires quite a bit of creative thinking, but there are certain procedures that can always be tried as first lines of attack. If nothing else, these will provide you with insights into what the puzzle really asks you to do.

Puzzle Properties

Puzzles in combinatory logic will ask you to discover or come up with some arrangement or combination of balls, colors, coins, and so on. For instance, you might be asked to determine how many draws are needed to get a pair of balls that match in color from a box of differently colored balls (example 1), or how many rounds must be played in a round-robin tennis tournament in order to determine a winner (example 2).

Example 1 The first illustrative puzzle requires that you reason carefully and not jump to conclusions about how to draw two items at random that match in color.

> In a box there are 20 balls, 10 white and 10 black. With a blindfold on, what is the least number you must draw out in order to get a pair of balls that matches?

Many newcomers to this kind of puzzle tend to reason somewhat along the following lines:

If the first ball that I pull out is white, then I will need another white one to match it. But the next ball might be black, as might be the one after that, and the one after that, and so on. So, in order to be sure that I get a match, I must remove all the black balls from the box—10 in all. The next one I remove after that will then necessarily be white. Including the first white ball I took out, the 10 black balls, and the one white ball that matches, 12 is the minimum number of balls I will need to draw out.

This line of reasoning, however, fails to take into account what the puzzle really requires the solver to do—namely, to match the color of *two* balls, not just the color of the first one drawn out, which happened to be *white*. The correct reasoning goes somewhat like this. Suppose the first ball you pull out is white. If you're lucky the next ball you draw out will also be white, and it's game over! But you cannot assume this best-case scenario. You must, on the contrary, assume the worst-case scenario, that is, that the next ball you pull out is black. Thus, after *two* draws, you will have taken out *one* white and *one* black ball from the box. Obviously, you could have pulled out a black ball first and a white one second. The end result would have been the same, *one* white and *one* black ball:

WORST-CASE SCENARIO

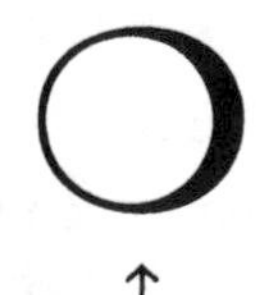

↑

FIRST DRAW = *WHITE* BALL

↑

SECOND DRAW = *BLACK* BALL

or

↑

FIRST DRAW = *BLACK* BALL

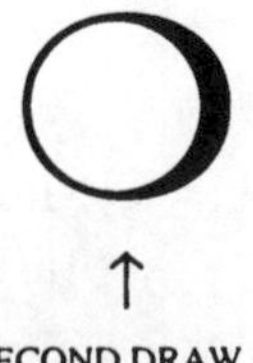

↑

SECOND DRAW = *WHITE* BALL

Now, the next ball you pull out from the box will be *either* white or black. Suppose you draw out a white ball; then it matches the white ball you pulled out first or second:

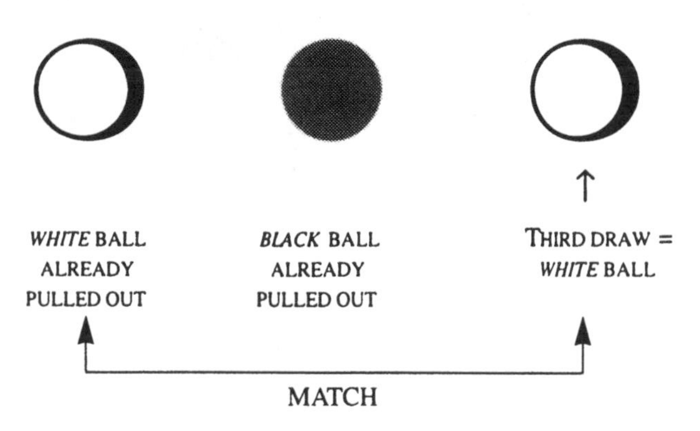

If, however, the next ball you draw out is black, then it matches the *black* ball you pulled out first or second:

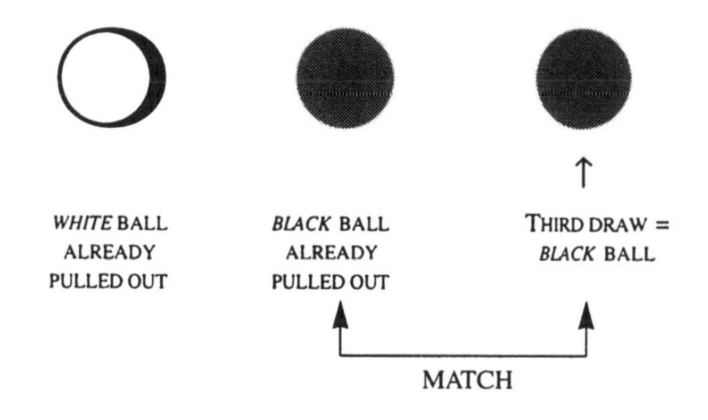

Therefore, no matter what color the *third* ball you draw out is, it will match the color of one of the two you had already pulled out. So, the least number of balls you will need to draw out from the box in order to get a pair of matching balls is *three.*

There are many versions of this puzzle, but the reasoning involved in solving them remains the same. Let's go through another version for the sake of illustration.

In a box there are 13 balls: 6 white, 4 black, and 3 with a star design. With a blindfold on, what is the least number you must draw out in order to get three matching balls?

Start by assuming a worst-case scenario: that is, assume that the *first three balls* you draw out will be of different colors: *one* white, *one* black, and *one* starred. As you discovered above, the order in which you draw the balls out is irrelevant. So, it is necessary to consider only, say, the following order of draws in your worst-case scenario:

WORST-CASE SCENARIO

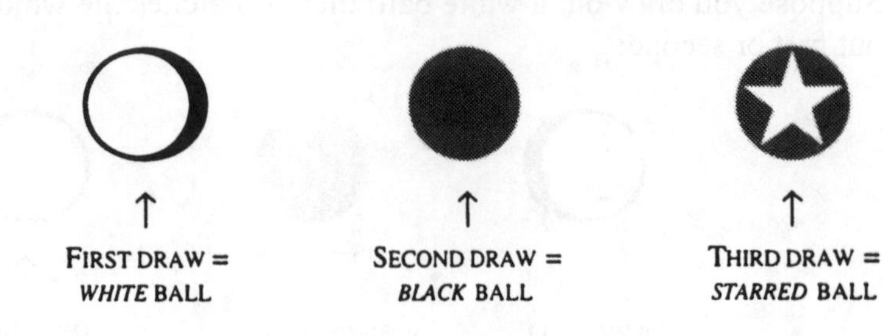

So far, you have made *three* draws. Now, suppose that for your *fourth* draw you get a white ball. You will then have a pair of white balls:

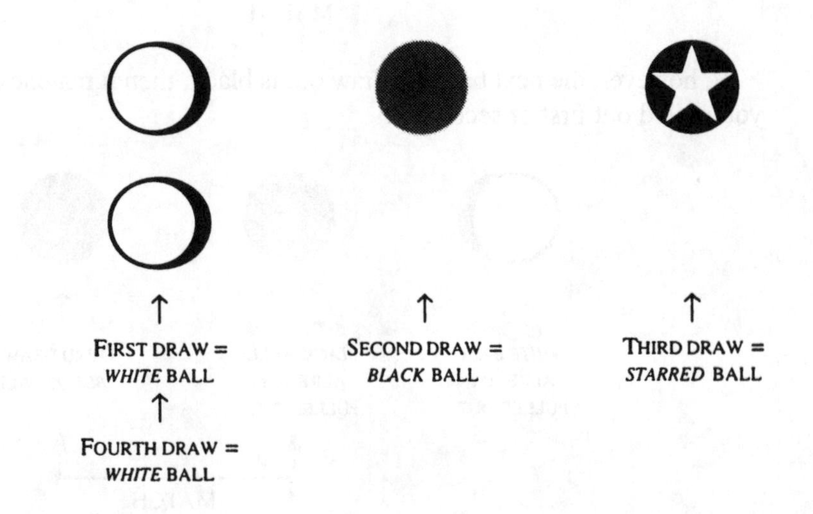

Again, for your *fifth* draw, you must assume a worst-case scenario: that is, you cannot assume that it will fortuitously produce a white ball. Suppose you draw out a black ball. Then your *fifth draw* produces a pair of black balls:

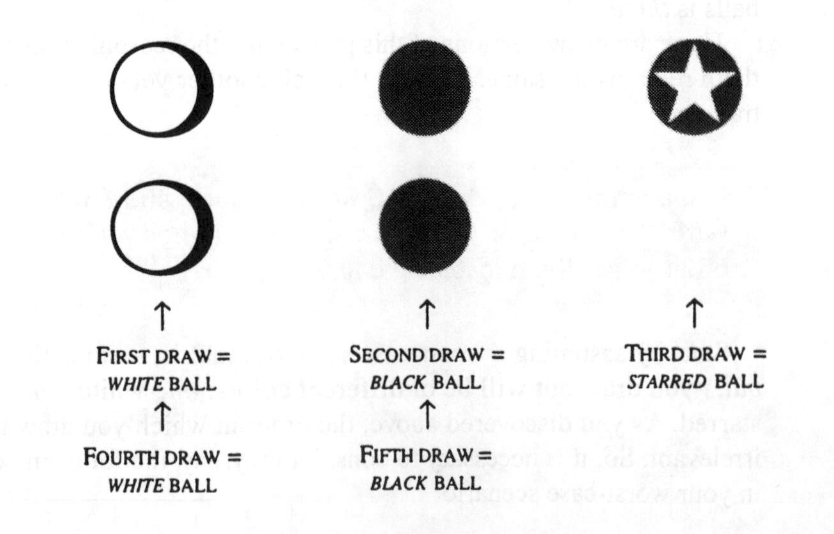

For your *sixth* draw you must again assume a worst-case scenario: namely, that it produces a *starred* ball. You will then have a pair of *starred balls* as well:

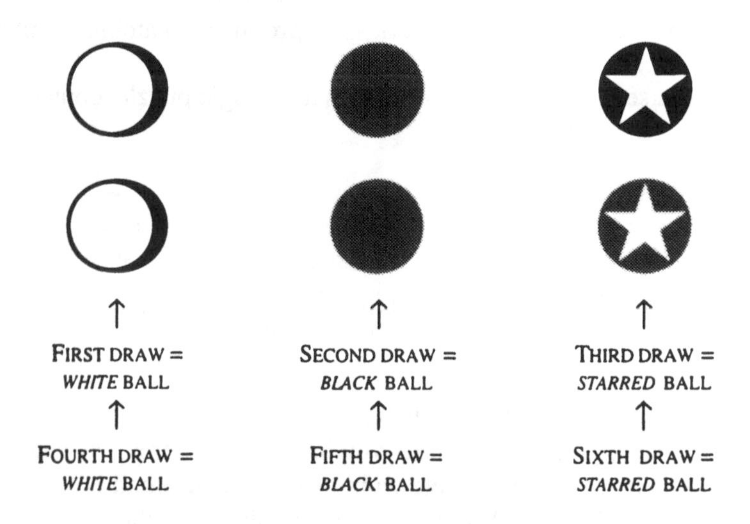

In any worst-case scenario you would end up with *three matching pairs* of balls after *six draws*. You can easily try out the other scenarios on your own: that is, let your *fourth draw* yield a black ball or a starred ball, and then follow the reasoning through in the same way.

Now, the next ball you draw out, be it white, black, or starred, will produce a matching triplet. For example, if you pull out a white one, then your *seventh draw* will produce a set of three white balls:

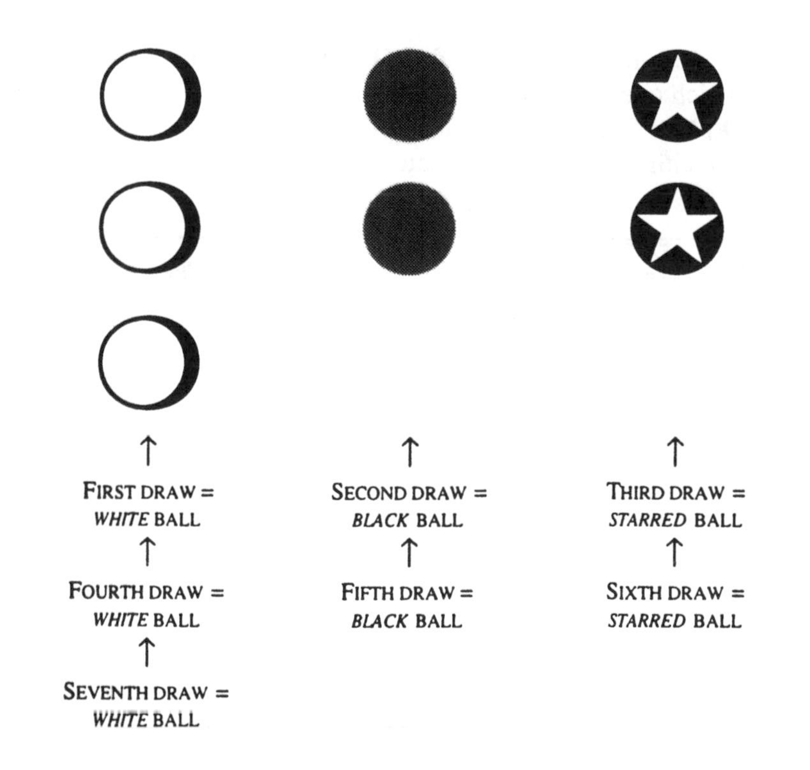

Of course, if you had pulled out a black one, then your *seventh draw* would have produced a set of *three black balls;* if you had pulled out a starred one, then your *seventh draw* would have produced a set of *three starred balls.* In conclusion, you will need *seven draws* to produce a matching triplet of balls.

Example 2 As a second example of a combinatory logic puzzle, consider the following simple "tournament" puzzle.

In a tennis club, five members decided to hold a round-robin tournament. The coach pointed out that since 5 is an odd number of players, one player should get a bye in the first round. How many rounds must be played in order to determine a winner?

For the sake of concreteness, give the five members letter names: **A, B, C, D, E.** Also, give **A** the bye. As you have done in previous chapters, the best way to keep track of the possible winners after each round is to set up a flow chart displaying the possible outcomes and results. Start the tournament by getting **B** and **C** to play against each other in the first round. Since either one can win, show the two possible outcomes in the chart as follows:

ROUND	SCENARIO 1	SCENARIO 2
1	**B** defeats **C** ↓ *Winner* = **B**	**C** defeats **B** ↓ *Winner* = **C**

Now, move on to round 2. **D** or **E** can be the next player. It doesn't matter which one it is. Choose **D**. Under scenario 1, **D** plays **B**, and under scenario 2, **D** plays **C**. **D** can either defeat or be defeated by **B** under scenario 1; or else **D** can either defeat or be defeated by **C** under scenario 2:

ROUND	SCENARIO 1	SCENARIO 2
1	**B** defeats **C** ↓ *Winner* = **B** ↓	**C** defeats **B** ↓ *Winner* = **C** ↓
2	**B** defeats **D** ↓ *Winner* = **B** *or* **D** defeats **B** ↓ *Winner* = **D**	**C** defeats **D** ↓ *Winner* = **C** *or* **D** defeats **C** ↓ *Winner* = **D**

What is the *end result* of the round? Under scenario 1, either **B** or **D** will go on to round 3; and under scenario 2, either **C** or **D** will go on to round 3. This might seem complicated, but it really isn't. All you really have to do is show the *end results* of the two scenarios clearly on the chart:

ROUND	SCENARIO 1	SCENARIO 2
1	**B** defeats **C** ↓ *Winner* = **B** ↓	**C** defeats **B** ↓ *Winner* = **C** ↓
2	**B** defeats **D** ↓ *Winner* = **B** *or* **D** defeats **B** ↓ *Winner* = **D** ↓ END RESULT ↓ *Winner* = **B** or **D**	**C** defeats **D** ↓ *Winner* = **C** *or* **D** defeats **C** ↓ *Winner* = **D** ↓ END RESULT ↓ *Winner* = **C** or **D**

Now, go on to round 3 with player **E**, the last player. Under scenario 1, **E** will play **B** or **D**. It really doesn't matter which one. Under scenario 2, **E** will play either **C** or **D**. Again, it doesn't matter which one. All you have to do is show all the possible outcomes for round 3 in your expanding flow chart:

ROUND	SCENARIO 1	SCENARIO 2
1	**B** defeats **C** ↓ *Winner* = **B** ↓	**C** defeats **B** ↓ *Winner* = **C** ↓
2	**B** defeats **D** ↓ *Winner* = **B** *or* **D** defeats **B** ↓ *Winner* = **D** ↓ END RESULT ↓ *Winner* = **B** *or* **D**	**C** defeats **D** ↓ *Winner* = **C** *or* **D** defeats **C** ↓ *Winner* = **D** ↓ END RESULT ↓ *Winner* = **C** *or* **D**

(continued)

(continued)

3	**B** *or* **D** defeats **E** ↓ *Winner* = **B** or **D** *or* **E** defeats **B** or **D** ↓ *Winner* = **E**	**C** or **D** defeats **E** ↓ *Winner* = **C** or **D** *or* **E** defeats **C** or **D** ↓ *Winner* = **E**

As you can now see, under scenario 1, **B** or **D** or **E** can win round 3; under scenario 2, **C** or **D** or **E** can win the round. Show these *end results* on your flow chart:

ROUND	SCENARIO 1	SCENARIO 2
1	**B** defeats **C** ↓ *Winner* = **B** ↓	**C** defeats **B** ↓ *Winner* = **C** ↓
2	**B** defeats **D** ↓ *Winner* = **B** *or* **D** defeats **B** ↓ *Winner* = **D** ↓ END RESULT ↓ *Winner* = **B** or **D**	**C** defeats **D** ↓ *Winner* = **C** *or* **D** defeats **C** ↓ *Winner* = **D** ↓ END RESULT ↓ *Winner* = **C** or **D**
3	**B** or **D** defeats **E** ↓ *Winner* = **B** or **D** *or* **E** defeats **B** or **D** ↓ *Winner* = **E** ↓ END RESULT ↓ *Winner* = **B** or **D** or **E**	**C** or **D** defeats **E** ↓ *Winner* = **C** or **D** *or* **E** defeats **C** or **D** ↓ *Winner* = **E** ↓ END RESULT ↓ *Winner* = **C** or **D** or **E**

The only person left is **A**, to whom you gave a bye at the start of the tournament. So, the winner of round 3 now plays A in round 4 to determine the overall winner of the tournament. Indicate the possible outcomes for the round as you have been doing so far:

ROUND	SCENARIO 1	SCENARIO 2
1	**B** defeats **C** ↓ *Winner* = **B** ↓	**C** defeats **B** ↓ *Winner* = **C** ↓
2	**B** defeats **D** ↓ *Winner* = **B** *or* **D** defeats **B** ↓ *Winner* = **D** ↓ END RESULT ↓ *Winner* = **B** or **D**	**C** defeats **D** ↓ *Winner* = **C** *or* **D** defeats **C** ↓ *Winner* = **D** ↓ END RESULT ↓ *Winner* = **C** or **D**
3	**B** or **D** defeats **E** ↓ *Winner* = **B** or **D** *or* **E** defeats **B** or **D** ↓ *Winner* = **E** ↓ END RESULT ↓ *Winner* = **B** or **D** or **E**	**C** or **D** defeats **E** ↓ *Winner* = **C** or **D** *or* **E** defeats **C** or **D** ↓ *Winner* = **E** ↓ END RESULT ↓ *Winner* = **C** or **D** or **E**
4	**B** or **D** or **E** defeats **A** ↓ *Winner* = **B** or **D** or **E** *or* **A** defeats **B** or **D** or **E** ↓ *Winner* = **A**	**C** or **D** or **E** defeats **A** ↓ *Winner* = **C** or **D** or **E** *or* **A** defeats **C** or **D** or **E** ↓ *Winner* = **A**

Now, you can see that under scenario 1, **B** or **D** or **E** or **A** can win round 4; and under scenario 2, **C** or **D** or **E** or **A** can win the round. Again, it doesn't matter who it is. All that really matters is to keep track of the end results:

Round	Scenario 1	Scenario 2
1	**B** defeats **C** ↓ *Winner* = **B** ↓	**C** defeats **B** ↓ *Winner* = **C** ↓
2	**B** defeats **D** ↓ *Winner* = **B** *or* **D** defeats **B** ↓ *Winner* = **D** ↓ END RESULT ↓ *Winner* = **B** or **D**	**C** defeats **D** ↓ *Winner* = **C** *or* **D** defeats **C** ↓ *Winner* = **D** ↓ END RESULT ↓ *Winner* = **C** or **D**
3	**B** or **D** defeats **E** ↓ *Winner* = **B** or **D** *or* **E** defeats **B** or **D** ↓ *Winner* = **E** ↓ END RESULT ↓ *Winner* = **B** or **D** or **E**	**C** or **D** defeats **E** ↓ *Winner* = **C** or **D** *or* **E** defeats **C** or **D** ↓ *Winner* = **E** ↓ END RESULT ↓ *Winner* = **C** or **D** or **E**
4	**B** or **D** or **E** defeats **A** ↓ *Winner* = **B** or **D** or **E** *or* **A** defeats **B** or **D** or **E** ↓ *Winner* = **A** ↓ END RESULT ↓ *Winner* = **B** or **D** or **E** or **A**	**C** or **D** or **E** defeats **A** ↓ *Winner* = **C** or **D** or **E** *or* **A** defeats **C** or **D** or **E** ↓ *Winner* = **A** ↓ END RESULT ↓ *Winner* = **C** or **D** or **E** or **A**

After this round a winner will emerge, even though you do not know who it will be. What you have discovered is that it will take *4 rounds* to determine a winner among the *5 players*. That is all the puzzle asked you to do.

Just for the sake of argument, how many *rounds* do you think it would take to determine a winner among *6 players*? Among *7 players*? Among *100 players*? Clearly, to construct a flow chart for 100 players would be unmanageable. So, it is obviously necessary to extrapolate some general rule or principle, if there is one. If you were to complete the appropriate flow charts for *6, 7,* and *8* players, you would soon discover that the *number of rounds* to be played is always *one less* than the *number of players* (a bye for one player is required for an *odd number of players* only). In algebraic notation, this can be phrased as follows: If there are n players ($n = 2, 3, 4, 5, \ldots$), then a total number of $n - 1$ rounds (*one less than the number of players*) will have to be played.

Number of Players	Number of Rounds
5	4
6	5
7	6
. . .	. . .
n	$n - 1$

Summary

In summary, solving puzzles in combinatory logic entails determining or coming up with some arrangement or combination under certain conditions. Here are basic lines of attack to keep in mind when solving such puzzles:

- Always assume a *worst-case scenario* in puzzles such as the one illustrated in example 1.
- Work out a solution for simpler versions of a puzzle, whenever possible. This will allow you to see if there is a general principle or rule in the simpler versions that can be used profitably. So, you might try out the required arrangement first with two or three balls, coins, and so on, rather than, say, twenty.
- Whenever possible, construct a visual aid (a diagram, a flow chart, etc.) that will show you the actual combinations, scenarios, outcomes, and so on that are required (as in example 2).

Puzzles 45–52

Answers, along with step-by-step solutions, can be found at the end of the chapter.

There are three closed boxes on a table that contain, separately, 10¢, 15¢, and 20¢ in nickels. However, they are labeled incorrectly. Someone takes the contents out of the box labeled 15¢, *two nickels,* and puts the nickels out in front of the box. Can you tell the contents of each box?

There are six checkers in a row on a table, three colored white and three colored black, with one space between the two sets (W = white, B = black):

Can you reverse the positions of the checkers by moving only one checker at a time?

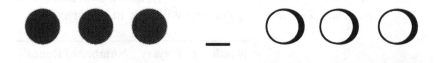

A checker may be moved over one adjacent checker into an empty space, or else, it may be moved one space into an empty space. You are not allowed to move a checker backward: that is, W's can only move to the right and B's to the left!

47 A traveler comes to a riverbank with a wolf, a goat, and a head of cabbage. There is only one boat for crossing, and it can carry no more than the traveler and one of the two animals or the cabbage. Unfortunately, if left alone together, the goat will eat the cabbage, and the wolf will eat the goat. How does the traveler transport his animals and his cabbage to the other side safely and soundly?

48 Imagine having three boxes: one with two black ties in it, the second one with two white ties in it, and the third one with one white tie and one black tie. The boxes are labeled, logically enough, **BB** (=two black ties), **WW** (=two white ties), and **BW** (=one black, one white tie). However, someone has switched the labels, so that now each box is labeled incorrectly. Taking out one tie at a time, can you determine the actual contents of each box in just one drawing?

49 Before they are blindfolded, three women are told that each one will have either a red or a blue cross painted on her forehead. When the blindfolds are removed, each person is then supposed to raise her hand if she sees a red cross and to drop her hand when she figures out the color of her own cross. Now, here's what actually happens. The three women are blindfolded and a red cross is drawn on each of their foreheads. The blindfolds are removed. After looking at each other, the three women raise their hands simultaneously. After a short time, one of the women lowers her hand and says, *"My cross is red."* How did she figure it out?

50 In a tennis club, 11 members decided to hold a round-robin tournament. The coach pointed out that since 11 is an odd number of players, one player should get a bye in the first round. How many games must be played in order to determine a winner?

51 There are three cups and ten coins on a table. You are asked to distribute the coins in the three cups according to the following two stipulations: (1) there must be no empty cup; (2) there must be seven coins in one cup and three in another. Is this possible?

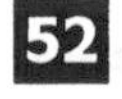

52 How many pets does my friend Morris have if all of them are dogs except two, all are cats except two, and all are rabbits except two?

Answers and Solutions

45 **Answer:** *The box labeled 10¢ actually contains four nickels (=20¢); the box labeled 15¢ actually contains two nickels (=10¢); and the box labeled 20¢ actually contains three nickels (=15¢).*

Solution: This puzzle requires you to match up box labels and their actual contents. First, list what you know: (1) the boxes contain 10¢ (=two nickels), 15¢ (=three nickels), and 20¢ (=four nickels); (2) each box is mislabeled: that is, if it says 10¢, then you know for certain that it does not have 10¢ in it, but 15¢ or 20¢; (3) the contents of the box labeled 15¢ are revealed to you—*two nickels* (10¢). Display these facts in visual form, labeling the boxes **A, B,** and **C**:

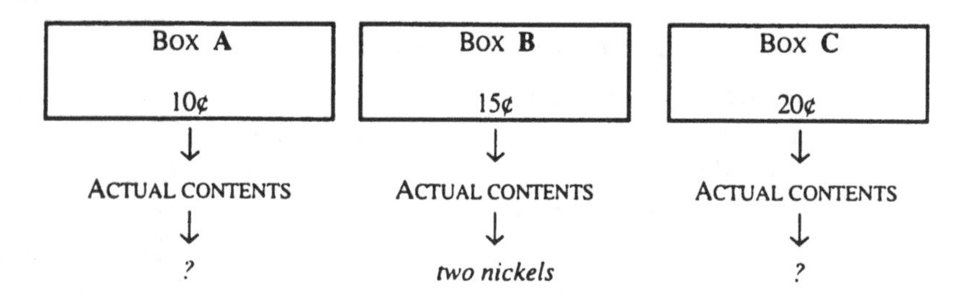

Now, you know that in the remaining two boxes, **A** and **C,** there are *three nickels* (15¢) and *four nickels* (20¢) in some order. This implies two possible scenarios:

Scenario 1

Box A	Box B	Box C
10¢	15¢	20¢
↓	↓	↓
Actual contents	Actual contents	Actual contents
↓	↓	↓
three nickels	*two nickels*	*four nickels*
↓	↓	↓
15¢	10¢	20¢

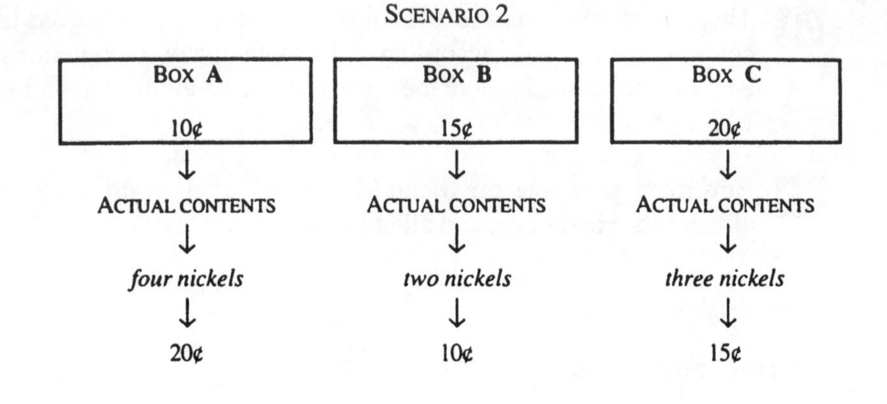

Scenario 1 contradicts a given fact. According to this scenario, **C** is labeled correctly as containing 20¢—contrary to the given fact that it should be labeled incorrectly. So, you must reject this scenario. Scenario 2, on the other hand, produces no contradictions: **A** contains 20¢, **B** contains 10¢, and **C** contains 15¢. So, the box labeled 10¢ actually contains four nickels (=20¢), the box labeled 15¢ actually contains two nickels (=10¢), and the box labeled 20¢ actually contains three nickels (=15¢).

46 **Answer:** *It can be done in 16 moves.*

Solution: It is advisable to start by solving simpler versions of this puzzle—that is, with two and four checkers instead of six. The two-checker version will be solved when you will have changed the set-up shown in **A** into the one shown in **B** (W = white, B = black), following the given rules of movement—(1) a checker may be moved over one adjacent checker into an empty space, or else, it may be moved one space into an empty space; (2) W's can only move to the right and B's to the left:

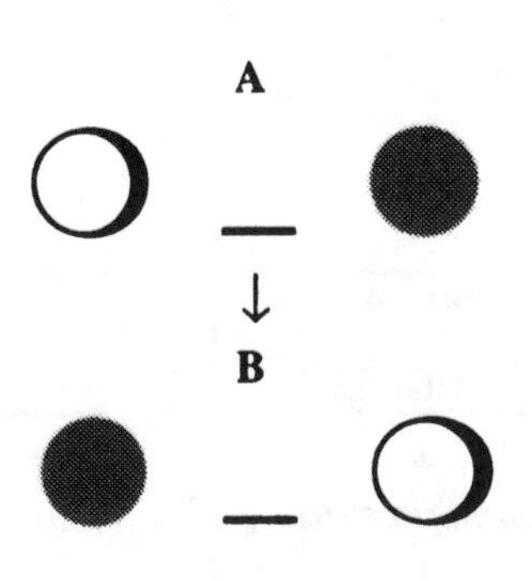

Keep track of your moves in sequence:

1. W _ B

Now, you can move the W over into the empty space, bringing W and B next to each other:

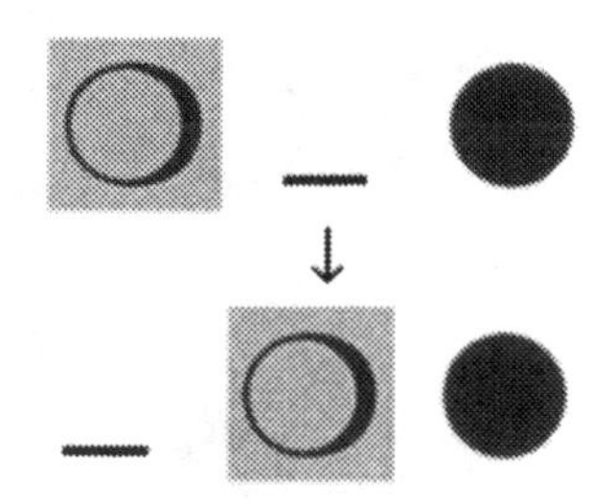

Note that you have created a new empty space by moving the W—namely, the space that it previously had occupied. The new arrangement can be represented as follows:

2. _ W B

Now, you can move the B over the W into the space to its left, creating a new space where the B used to be:

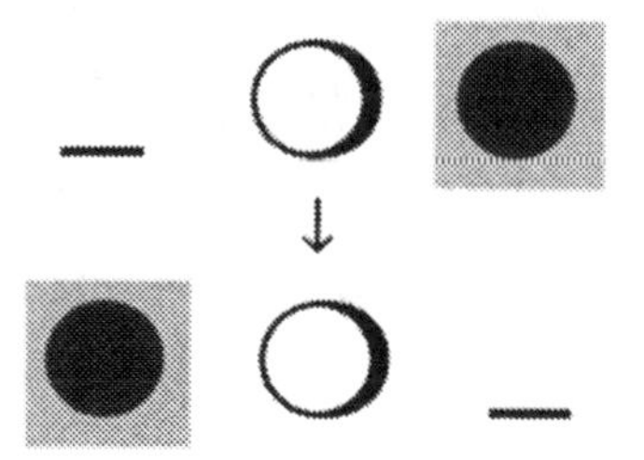

This is the third arrangement:

3. B W _

Finally, you can move the W over into the empty space, leaving a space where the W used to be:

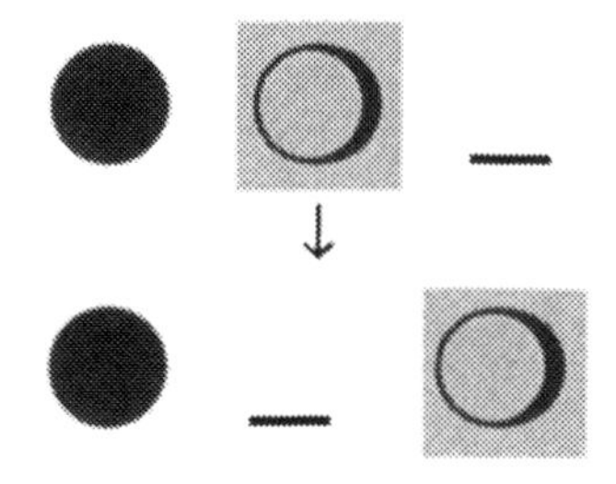

This is the required arrangement:

4. B _ W

Now, turn your attention to solving the four-checker version of the puzzle:

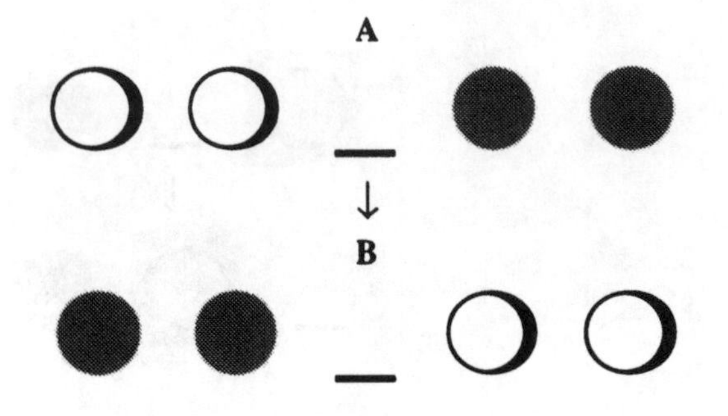

Once again, you can show your initial arrangement with letters as follows:

1. W W _ B B

Start by moving the W into the empty space next to a B, creating a new space where the W used to be:

2. W _ W B B

Now, move the B over the W into the empty space:

3. W B W _ B

Next, move the rightmost B over into the empty space:

4. W B W B _

After that, move the W over the B into the empty space:

5. W B _ B W

Now, you can put the leftmost W into the empty space by moving it over the B:

6. _ B W B W

This leaves an empty space into which you can move the leftmost B:

7. B _ W B W

You can now move the rightmost B over the W into the empty space:

8. B B W _ W

Finally, you can move the W into the empty space to complete the solution:

9. B B _ W W

As a general rule, note that the solution seems to depend on alternating the combination W B W B wherever possible. Solving the six-checker version of this puzzle should now cause you little difficulty. Here is the sequence of arrangements summarized for you without commentary:

1.	W	W	W	_	B	B	B
2.	W	W	_	W	B	B	B
3.	W	W	B	W	_	B	B
4.	W	W	B	W	B	_	B
5.	W	W	B	_	B	W	B
6.	W	_	B	W	B	W	B
7.	_	W	B	W	B	W	B
8.	B	W	_	W	B	W	B
9.	B	W	B	W	_	W	B
10.	B	W	B	W	B	W	_
11.	B	W	B	W	B	_	W
12.	B	W	B	_	B	W	W
13.	B	_	B	W	B	W	W
14.	B	B	_	W	B	W	W
15.	B	B	B	W	_	W	W
16.	B	B	B	_	W	W	W

47 **Answer:** *The traveler can do it in eight trips back and forth.*

Solution: This is a version of a puzzle that was originally invented by a friend of Charlemagne—the English scholar of the late eighth century, Alcuin.

Start by showing the initial situations on both sides of the crossing, before the traveler starts ferrying back and forth (W = wolf, G = goat, C = cabbage):

Original Side	Other Side
1. W G C	_ _ _

The traveler can start by transporting the goat over on the boat—recall that he can only take himself and one animal or the cabbage with him. This leaves the wolf and the cabbage alone without any problems, because the wolf does not eat cabbage:

Original Side		Other Side
1. W G C		_ _ _
2. W _ C	→	_ G _

The traveler can then come back alone:

Original Side		Other Side
1. W G C		_ _ _
2. W _ C	→	_ G _
3. W _ C	←	_ G _

From the original side, he can then take the wolf across with him:

Original Side		Other Side
1. W G C		_ _ _
2. W _ C	→	_ G _
3. W _ C	←	_ G _
4. _ _ C	→	W G _

But to go back and get the cabbage on the original side, the traveler cannot, clearly, leave the wolf and goat alone, for the former would eat the latter. What to do? Well, he could bring the goat back for the ride, leaving the wolf harmlessly by itself. Actually, the reason for taking the goat back at this point will soon become evident:

Original Side		Other Side
1. W G C		_ _ _
2. W _ C	→	_ G _
3. W _ C	←	_ G _
4. _ _ C	→	W G _
5. _ G C	←	W _ _

Back on the original side, he can leave the goat and take the cabbage with him over to the wolf, who does not eat cabbage. This is why the astute traveler knew enough to go back on the previous trip with the goat and not the wolf!

Original Side		Other Side
1. W G C		_ _ _
2. W _ C	→	_ G _
3. W _ C	←	_ G _
4. _ _ C	→	W G _
5. _ G C	←	W _ _
6. _ G _	→	W _ C

The traveler can then go back alone, leaving the wolf and the cabbage on the other side with no disastrous consequences.

	Original Side		Other Side
1.	W G C		_ _ _
2.	W _ C	→	_ G _
3.	W _ C	←	_ G _
4.	_ _ C	→	W G _
5.	_ G C	←	W _ _
6.	_ G _	→	W _ C
7.	_ G _	←	W _ C

For his last trip, he brings the goat over and then continues happily on his journey.

	Original Side		Other Side
1.	W G C		_ _ _
2.	W _ C	→	_ G _
3.	W _ C	←	_ G _
4.	_ _ C	→	W G _
5.	_ G C	←	W _ _
6.	_ G _	→	W _ C
7.	_ G _	←	W _ C
8.	_ _ _	→	W G C

48 **Answer:** *The actual contents of each box can be determined in just one drawing if that drawing is made from the box labeled* ***BW****.*

Solution: The first thing to do is to display the given facts in visual form, labeling the boxes as 1 = **BB** (two black ties), 2 = **WW** (two white ties), and 3 = **BW** (a black and a white tie). A box will not contain the combination that its label says; but it could contain any one of the other two combinations. Indicate this as *possible contents:*

1	2	3
BB	**WW**	**BW**
↓	↓	↓
Possible contents	Possible contents	Possible contents
↓	↓	↓
WW *or* **BW**	**BB** *or* **BW**	**WW** *or* **BB**

Now, start by drawing from box 1—the one labeled incorrectly **BB**:

➤ **Scenario 1: Drawing from box 1**

- □ As you can see from its possible contents of **BW** and **WW**, if you draw a **B**, then box 1 contains **BW**. If you draw a **W**, then the box contains either **BW** or **WW**. As always, you must assume the latter worst-case scenario.
- □ Your second draw will, of course, tell you which combination it actually contains. You can then determine the actual contents of the other two boxes without drawing any ties from them.
- □ Assume that you drew **BW** from box 1. Now look at the possible contents of boxes 2 and 3. Clearly, by the process of elimination, box 2 contains **BB** (not **BW**) and, therefore, box 3 contains **WW**. If instead you drew **WW**, then by the process of elimination, box 3 contains **BB** (not **WW**) and, therefore, box 2 contains **BW**.
- □ So, if you draw from box 1, you will need two drawings to be able to infer what is in each box.

Now, draw your tie from box 2 instead—the one incorrectly labeled **WW**:

➤ **Scenario 2: Drawing from box 2**

- □ As you can see from its possible contents of **BB** and **BW**, if you draw a **W**, then that box contains **BW**. If you draw a **B**, then the box contains either **BW** or **BB**. As before, you must assume the latter worst-case scenario.
- □ Your second draw will, of course, tell you which combination it actually contains. You can then deduce what the other boxes contain without drawing any ties from them.
- □ Assume that you drew **BW** from box 2. Now, look at the possible contents of boxes 1 and 3. Clearly, by the process of elimination, box 1 contains **WW** (not **BW**) and, therefore, box 3 contains **BB**. If instead you drew **BB**, then by the process of elimination, box 3 contains **WW** (not **BB**) and, therefore, box 1 contains **BW**.
- □ Once again, if you draw from box 2, you will need two drawings to be able to infer what is in each box.

Let's see what happens if you draw your tie from box 3.

➤ **Scenario 3: Drawing from box 3**

- □ As you can see from its possible contents of **BB** and **WW**, if you draw a **B**, then that box contains **BB**. If you draw a **W**, then the box contains **WW**.

- A second drawing is therefore unnecessary, since you know that it will produce a match—either **BB** or **WW**. So, in one drawing you will know what is in box 3. You can also deduce what the other boxes contain without drawing any ties from them.
- Assume that you drew **B** from box 3. It therefore contains **BB**. Now, look at the possible contents of boxes 1 and 2. Clearly, by the process of elimination, box 2 contains **BW** (not **BB**) and, therefore, box 1 contains **WW**. If instead you drew **W** from box 3, then you know that it contains **WW**. Then, by the process of elimination, box 1 contains **BW** (not **WW**) and, therefore, box 2 contains **BB**.

In conclusion, you can determine the actual contents of each box in one drawing if you make your draw from the box labeled **BW**.

49 **Answer:** *One of the women reasons as follows: The other two cannot determine their color. This means that I too have a red cross.*

Solution: This is a classic puzzle in the genre of combinatory puzzles. Call the three women **A**, **B**, and **C**. Assume that **A** is the one who figured out the color of the cross on her head.

Consider **A**'s initial reaction. She looks at **B** and **C** and notices that they both have a red cross. So, naturally, she will put up her hand as she has been instructed to do. Similarly, **B** also sees two red crosses. So, she too raises her hand. **C** also sees two red crosses, and of course she too will raise her hand. So, how did **A** figure it out? **A** must have had a flash of insight, thinking as follows:

Let me assume that I have a blue cross on my forehead. If that is so, then one of the other two, say **B**, would know that she doesn't have a blue cross because otherwise **C**, seeing two blue crosses—mine and **B**'s—would not have put up her hand. But this has not occurred. So, **B** and **C** cannot determine their color. This means that I too have a red cross.

---

50 **Answer:** *10 rounds.*

Solution: This puzzle is exactly like example 2 above, so it is not necessary to go into the full solution here. Only the main insight gained by solving the previous puzzle will be used.

Recall that the *number of rounds* to be played is always *one less* than the *number of players* (a bye for one player is required for an *odd number of players* only). In algebraic notation, if there are n players ($n = 2, 3, 4, 5, \ldots$), then a total number of $n - 1$ rounds (one less than the number of players) will have to be played.

Number of Players	Number of Rounds
5	4
6	5
7	6
...	...
n	$n - 1$

In this case, $n = 11$, so the number of rounds to be played, $n - 1$, is 10.

51 **Answer:** *By putting one cup inside another, the same set of coins can belong to more than one cup.*

Solution: This puzzle seems to be unsolvable. Note, however, that the puzzle does not prohibit you from putting one cup inside another. Indeed, by doing so, the same set of coins can belong to more than one cup. So, let's see if this line of attack will work.

Put the ten coins into the three cups separately as follows:

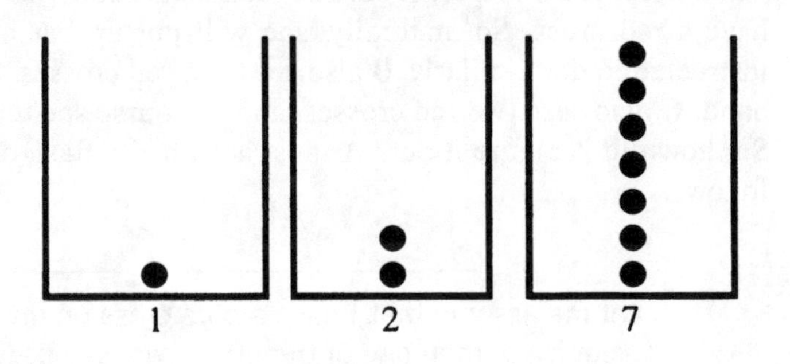

Now, insert the cup with two coins inside it into the one with one coin inside it:

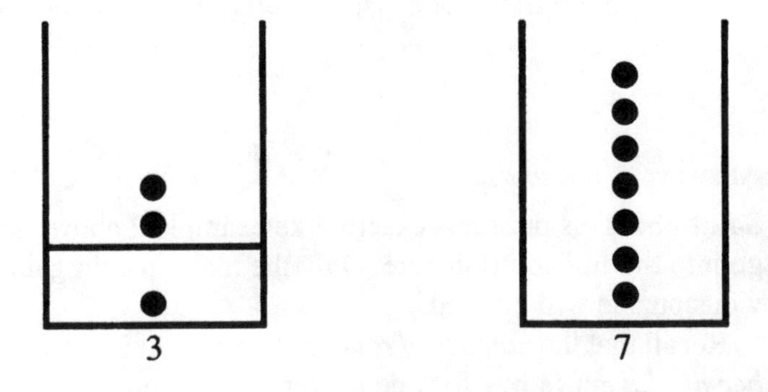

You have now satisfied the puzzle's two stipulations—(1) there is no empty cup; and (2) there are seven coins in one cup and three in another.

52 **Answer:** *One dog, one cat, and one rabbit.*

Solution: Start by representing the actual number of dogs Morris has with d, the actual number of cats with c, and the actual number of rabbits with r. Now, convert this combinatory puzzle into an algebraic problem.

You are told that the total number of pets Morris has—$d + c + r$—minus 2 equals the actual number of dogs he has:

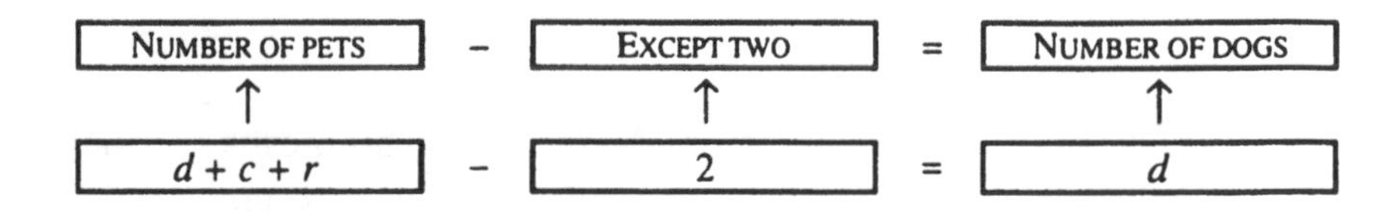

Simplifying the equation, $d + c + r - 2 = d$, you'll get:

$$(1) \quad c + r = 2$$

You are also told that his total number of pets minus 2 equals the actual number of cats he has:

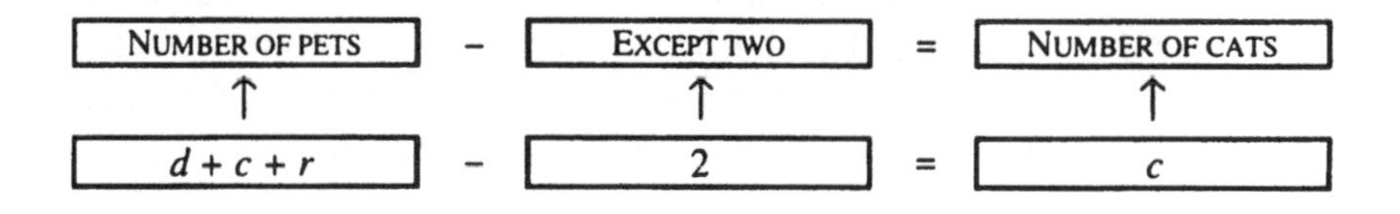

Simplifying the equation $d + c + r - 2 = c$, you'll get:

$$(2) \quad d + r = 2$$

Finally, you are told that his total number of pets minus 2 is equal to the actual number of rabbits he has:

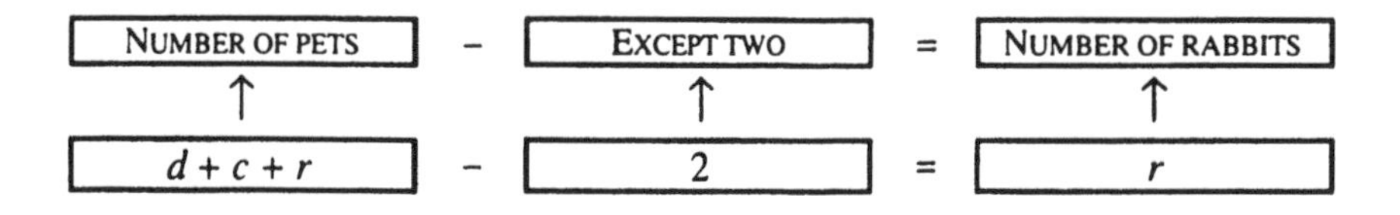

Simplifying the equation $d + c + r - 2 = r$, you'll get:

$$(3) \quad d + c = 2$$

Put the three equations together:

$$(1) \quad c + r = 2$$

$$(2) \quad d + r = 2$$

$$(3) \quad d + c = 2$$

From Equations (1) and (2), it can be seen that:

$$c + r = d + r$$

or

$$c = d$$

And from Equations (2) and (3), it can be seen that:

$$d + r = d + c$$

or

$$c = r$$

Since $c = d$, and $c = r$, then, clearly, $d = r$. This means, obviously, that $d = c = r$. So, Morris has the same number of dogs, cats, and rabbits. Try a few combinations. If Morris has 6 pets in all, then he has 2 dogs, 2 cats, and 2 rabbits. However, this combination does not work out, since all his pets except 2, or 6 – 2, are supposed to be dogs. Thus, he is supposed to have 4 dogs, not 2! The same contradictory finding applies to the cats and to the rabbits.

Try a 9-pet combination of 3 dogs, 3 cats, and 3 rabbits. This also does not work out because the number of dogs he is supposed to have is 9 – 2 or 7, not 3! The same contradictory finding applies to the cats and to the rabbits.

But we missed the simplest case of all. If Morris has just 3 pets, then he has 1 dog, 1 cat, and 1 rabbit. This, of course, works out perfectly, since all his pets except 2, namely 3 – 2 or 1, are supposed to be dogs. And, indeed, he has 1 dog. Similarly, he has 3 – 2 or 1 cat, and 3 – 2 or 1 rabbit.

7 Puzzles in Geometrical Logic

Geometry is one of the oldest branches of mathematics. The term derives from the ancient Greek *geo* "earth" and *metrein* "to measure." This is an accurate description of the works of the early geometers, who were concerned with such problems as measuring the size of fields and laying out accurate right angles for the corners of buildings. This type of practical geometry, which flourished in ancient Egypt, Sumer, and Babylon, was refined and systematized by the Greeks. In the sixth century B.C. the Greek mathematician Pythagoras laid the cornerstone of scientific geometry by transforming it into a full-fledged mathematical discipline. Pythagoras admired geometry not only on account of its great practicality but also because of its tremendous intellectual value and beauty.

As you might imagine, the selection of puzzles included in this chapter is not an extensive one. Any claim to thoroughness would require several tomes, not the present selection. The solution strategies illustrated in this chapter are nevertheless applicable to the solution of a large number of geometrical puzzles.

How To...

Solving puzzles in geometrical logic requires quite a bit of imagination, but there are several approaches that can always be tried as preliminary lines of attack. If nothing else, these will allow you to understand better what the puzzle asks you to do.

Puzzle Properties

The most common type of geometrical puzzle, which is illustrated in example 1, consists of drawing or altering a given figure according to some specific instruction or set of instructions. Puzzles of this type are commonly solved by a trial-and-error approach. If you do not see how to solve the puzzle from the very outset, try anything that comes to mind. You never know. It might just work out! After several unsuccessful attempts, you should start suspecting that your approach is probably the wrong one. Check out hunches and other approaches, one by one. Sooner or later, something might work!

The second most common type of geometrical puzzle is illustrated in example 2 below. This requires some knowledge of the fundamentals of geometry, such as the fact that the area of a rectangle is equal to its length multiplied by its width, that the square on the hypotenuse of a right-angled triangle equals the sum of the squares on the other two sides, and so on.

Example 1 The *nine-dot four-line* puzzle is a classic in the geometrical puzzle genre.

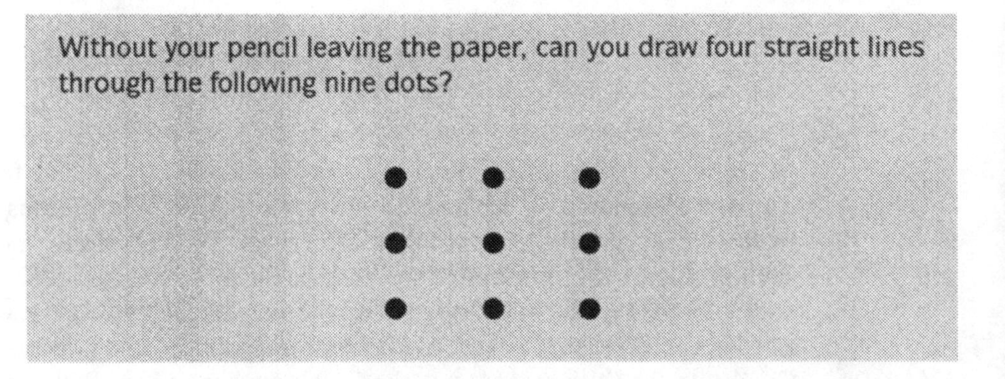

Without your pencil leaving the paper, can you draw four straight lines through the following nine dots?

Most people will approach this puzzle at first by attempting to join the dots as if they were located on the *perimeter* (boundary) of a square:

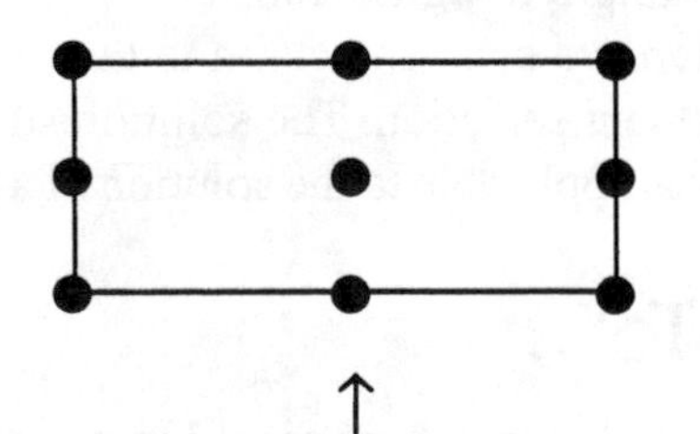

↑
NINE DOTS SEEN AS BEING ON THE PERIMETER OF A SQUARE

But this line of reasoning does not yield a solution:

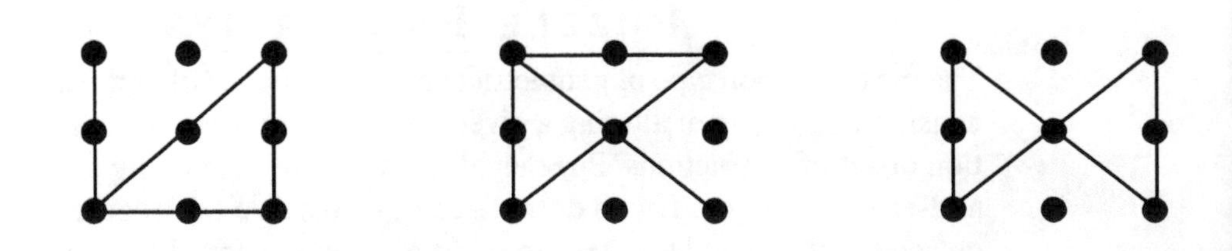

So, it would appear to be an impossible task to join all of the nine dots with four lines, since one or two dots seem always to be left over. Or is it? It is certainly impossible to join the four dots with four lines, if you view the dots as *being on the perimeter of a square.*

Let's try another approach. Put your pencil on, say, the dot on the bottom right, joining it to the dots that make up a diagonal going 45° to the left. (You could start with any of the four corner dots, the reasoning would be the same.)

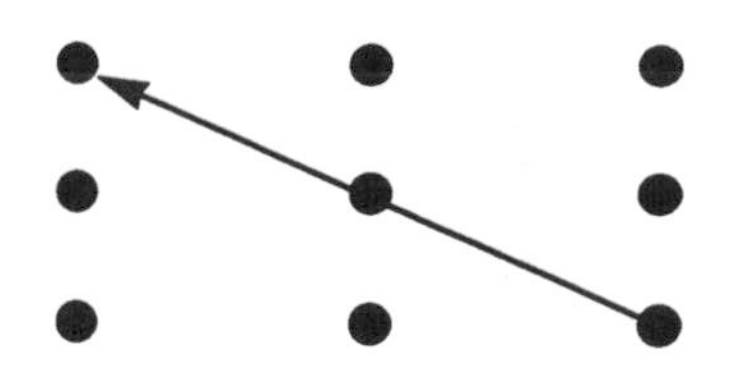

Now, with a second line turn right, and join the two dots along the top, *without stopping at the end dot.* In fact, go a bit beyond that dot, so that you can be in line with the two dots that fall on a diagonal going down 45° to the left:

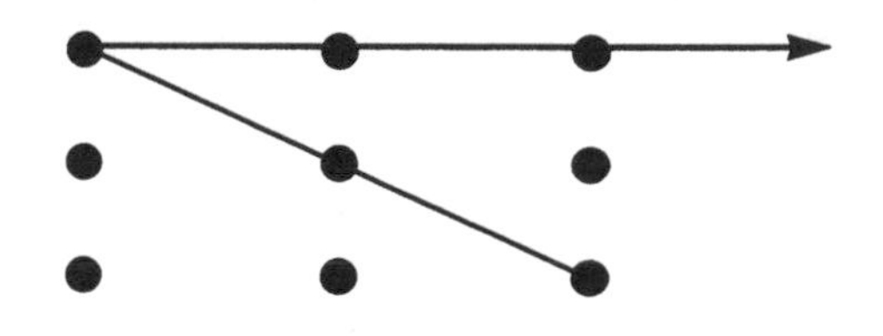

Your third line can now be drawn as that very diagonal. Once again, go a bit beyond the end dot, so that you can be in line with the remaining two dots. Indeed, consider the dots as part of a line going straight up:

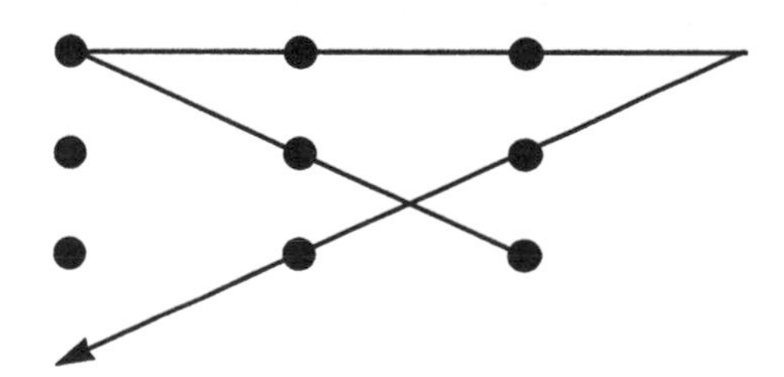

Finally, with a fourth line, join the last two dots.

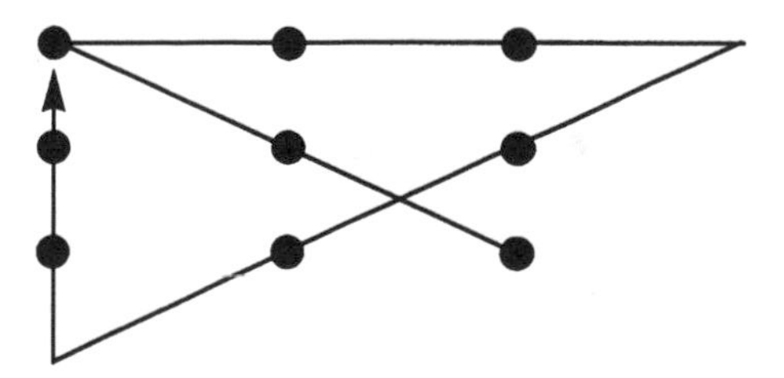

The solution is now complete. Reflecting on how this puzzle came to be solved is instructive. At the outset, an approach was tried out that didn't seem to be working out. So, we started to suspect that it was impossible to solve the puzzle in that way. The proper analysis of why this was so—namely, that the dots were not to be perceived as being on the perimeter of a square—was the key that opened the door to the solution.

Example 2 The second type of geometrical puzzle requires some knowledge of basic school geometry and of some elementary algebra. This type is invariably solved with the aid of pictorial representation:

> My neighbor built a rectangular fish pool 4 feet wide by 9 feet long. He also built a walk of uniform width around it. He can only afford to make the area of the walk 68 square feet. So, how wide should he make his walk?

Without a suitable diagram to help you visualize the walk, it is very difficult to solve this puzzle. The puzzle states that the fish pool is *rectangular.* So, the first thing to do is to translate that statement into an appropriate figure:

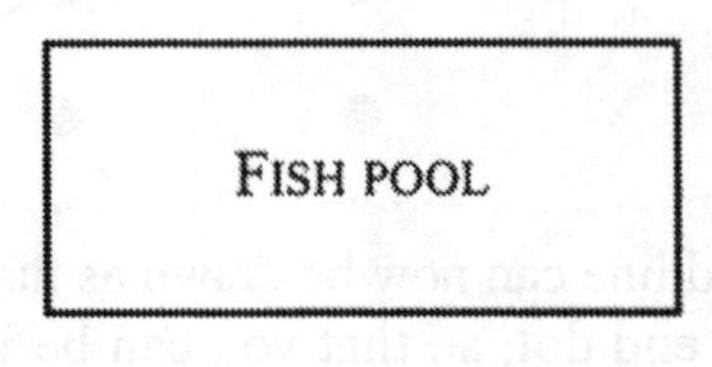

The puzzle also tells you what the dimensions of the pool are: 9 feet by 4 feet. Put this information on the figure:

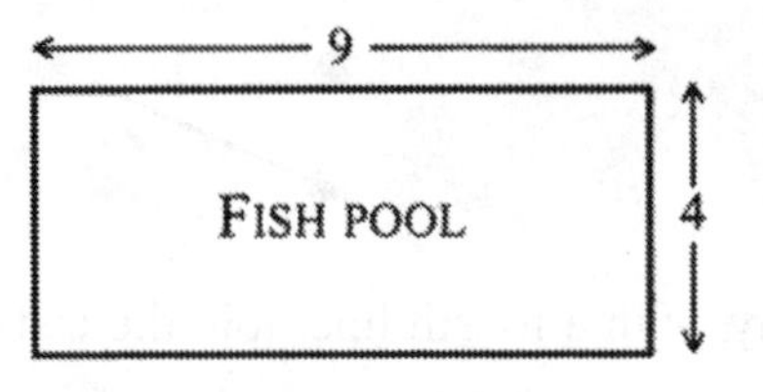

Now, draw the walk around the fish pool:

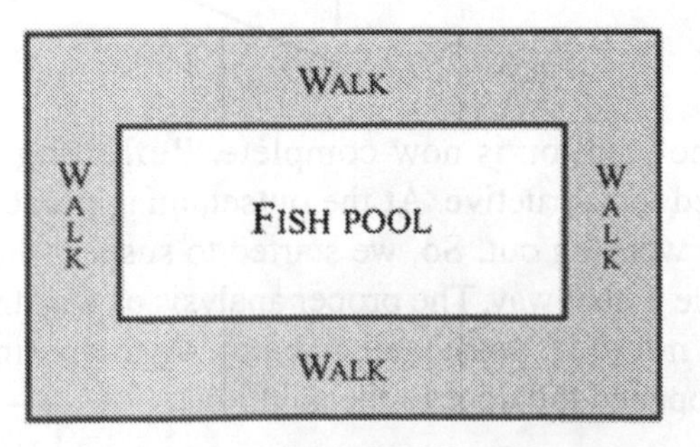

You are told that the walk is of uniform width all the way around the pool. What does that imply? It means, of course, that its width from the pool is always the same. Represent this width with x:

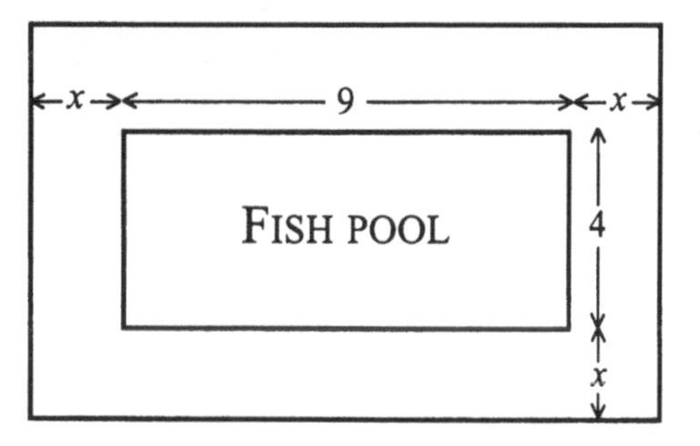

The diagram now allows you to determine easily the dimensions of the larger rectangle, that is, of the rectangle formed by the pool and the walk together:

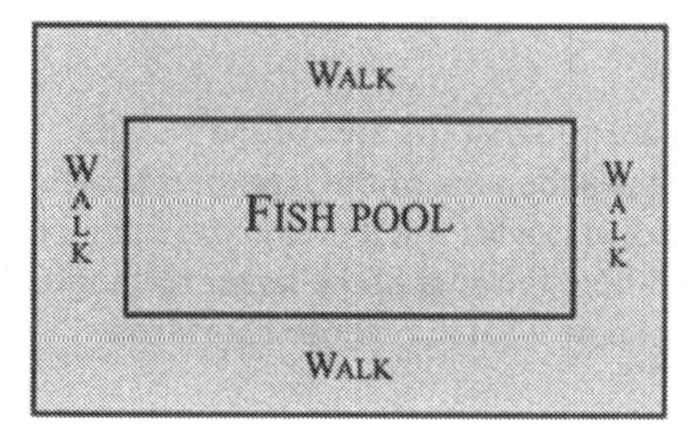

The length of this rectangle is, of course, $x + 9 + x$, or $2x + 9$ feet:

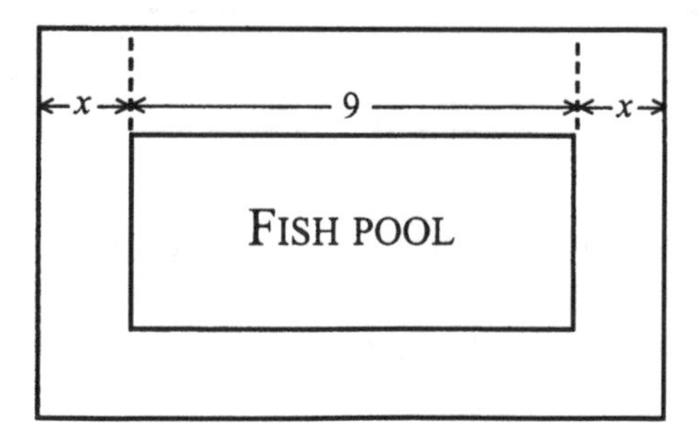

Its width is $x + 4 + x$, or $2x + 4$ feet:

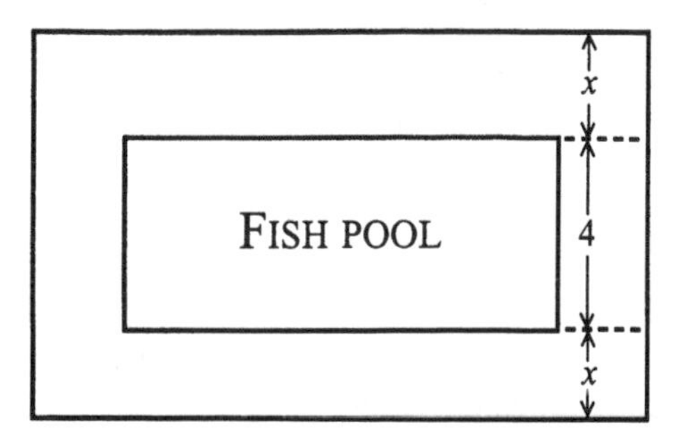

Now that you know the length and width of the larger rectangle, you can easily compute its area:

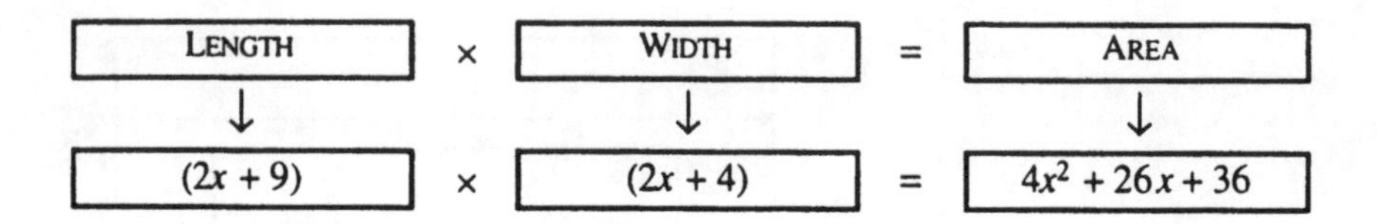

Next, compute the area of the fish pool:

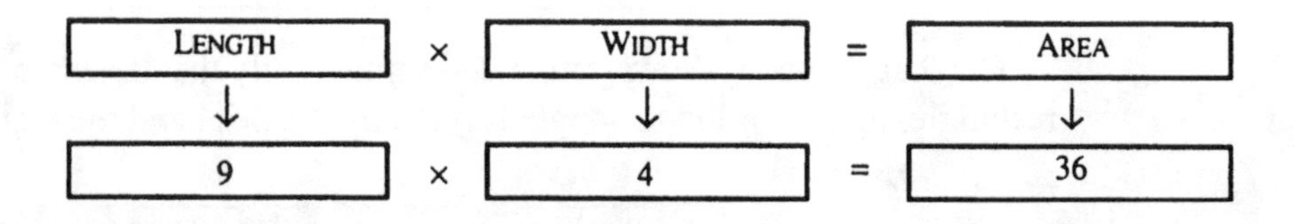

You are told that the area of the walk is to be 68 square feet. The area of the walk equals the area of the larger rectangle minus the area of the fish pool—because that is what is left over if you subtract the fish pool area—the smaller rectangle—from the area of the larger rectangle. From the calculations above, you already know what the relevant areas are:

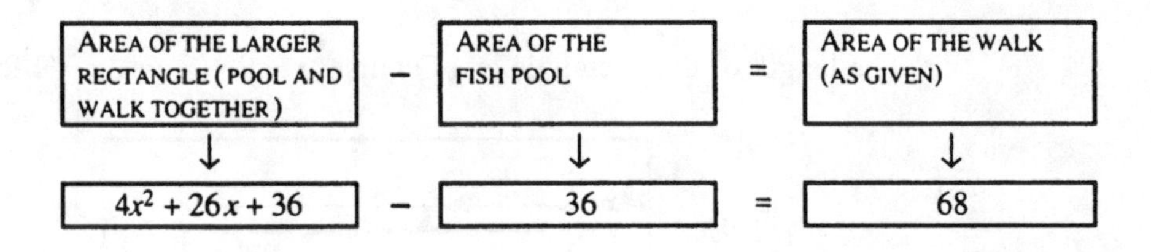

If you have forgotten how to solve for x in the equation $4x^2 + 26x + 36 - 36 = 68$, or $4x^2 + 26x = 68$, you should review your school notes on quadratic equations. Without going into the technical details here, the answer is $x = 2$. So, the neighbor should make his walk 2 feet wide.

Summary

The type of puzzle illustrated in example 1 cannot be approached systematically in the same way that, say, puzzles in logical deduction, truth logic, or algebraic logic can. Always try out an initial approach (connecting dots, dissecting figures, etc.). If it works, end of matter. If it does not, then analyze why it has turned out to be fruitless. Try a new hunch.

In the case of puzzles such as the one illustrated in example 2, the key to a solution is drawing a simple diagram showing in pictorial form everything that the puzzle states in words.

 Puzzles 53–57

Answers, along with step-by-step solutions, can be found at the end of the chapter.

53 Given the dimensions of the radius in inches, can you calculate the length of the rectangle's diagonal that goes from **A** to **B**?

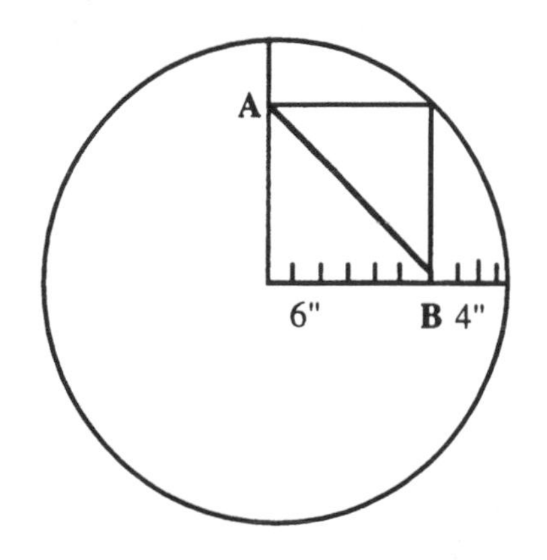

54 A farmer wants to fence in his chicken pen with 60 feet of wire, using the side of his barn for one side of the pen. If the length of his pen is to be three times its width, what are its dimensions?

55 The length of a rectangular sheet of cardboard is twice its width. In each corner, a 2-inch square is cut out so that the sides can be turned up to form a box. If the box has a volume of 60 cubic inches, what were the dimensions of the original piece of cardboard?

56 How can the following rectangle, with its two tabs, be cut into two pieces to make a complete rectangle?

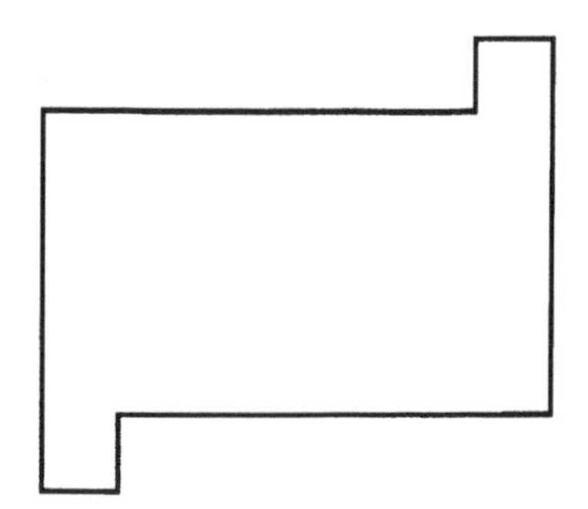

57 With one cut, you will slice a pie into two pieces. With a second cut that crosses the first one, you will produce four pieces. With a third cut, you can produce as many as seven pieces:

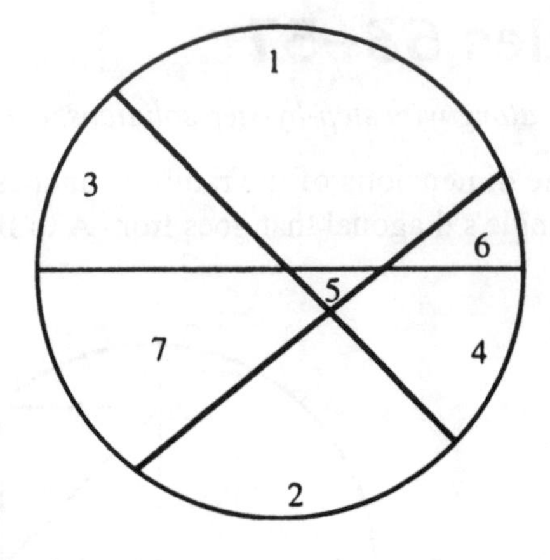

What is the largest number of pieces you can get with six straight cuts?

Answers and Solutions

 **Answer:** *10 inches.*

Solution: You will find this puzzle virtually impossible to solve unless you do something that is actually quite simple, but not totally obvious. Recall from your school geometry that the two diagonals of a rectangle equal each other in length. So, go ahead and draw the other diagonal:

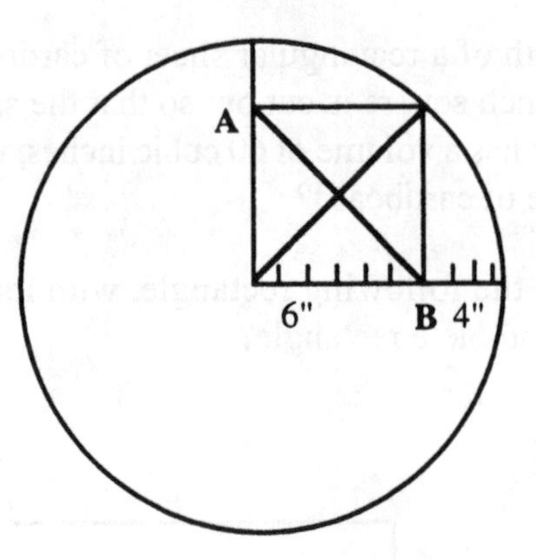

Now, do you see the solution? The diagonal that has just been drawn is, in effect, a radius of the circle. It can be seen from the illustration that the radius is equal to 6 inches plus 4 inches, or 10 inches in total. All radii of a circle are equal. So, the diagonal that was just drawn is also 10 inches long. And since the diagonals of a rectangle are equal, the length of the diagonal **AB** is also 10 inches.

54 **Answer:** *Width = 12 feet, length = 36 feet.*

Solution: First, draw the barn and the chicken pen, showing the width of the pen as x and its length as $3x$—given that the length is to be three times the width:

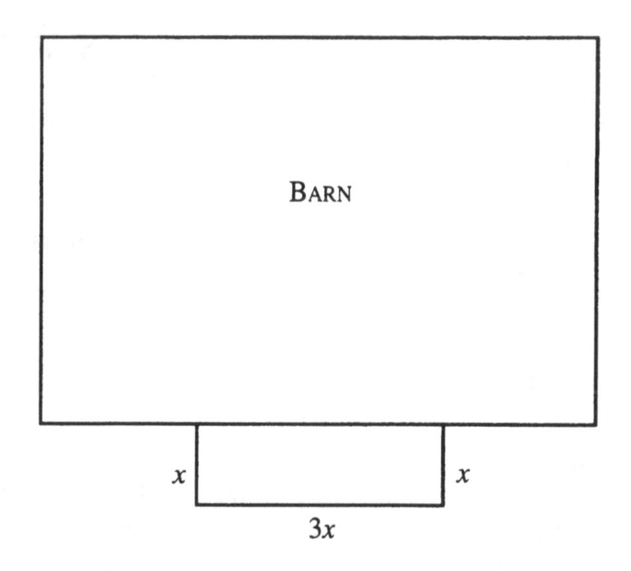

You are told that the farmer has 60 feet of wire to cover the three sides of the pen. The three sides add up, clearly, to $x + x + 3x$, or $5x$ feet in total. This total equals 60 feet; so, $5x = 60$. Solving for x, you'll get: $x = 12$. In conclusion, the dimensions of the pen are as follows: width = 12 feet; length = 36 feet ($=3x$).

55 **Answer:** *7 inches by 14 inches.*

Solution: Draw the rectangular piece of cardboard. On it, let x stand for its width and $2x$ for its length (because it is twice its width); also show 2-inch square cut-outs from each of its corners:

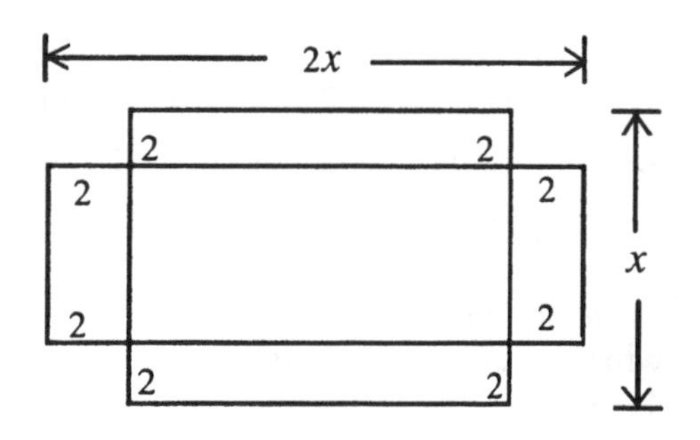

Now, consider the box that is formed by turning up the corners of the cardboard. As can be easily seen, the width of the box that is formed is $x - 4$ inches (the width of the original rectangle minus the two 2-inch cuts), and its length is $2x - 4$ inches (the length of the original rectangle minus the two 2-inch cuts). The height of the box is, of course, 2 inches.

Recall from your school days that the volume of a rectangular container is length × width × height. You are told that the volume of the box is equal to 60 cubic inches:

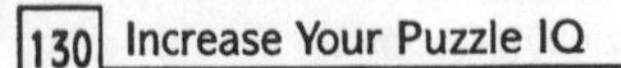

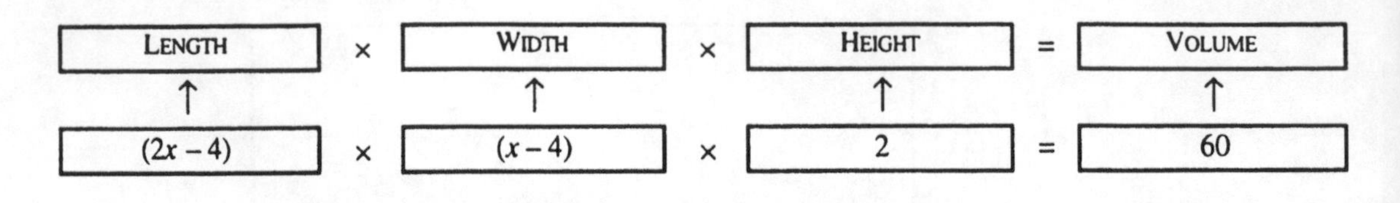

Solving for x, you'll get: $x = 7$. So, the dimensions of the original piece of cardboard were 7 inches by 14 inches.

56 **Answer:** *Cut the rectangle in a zigzag fashion with each cut equal to the length of the tab.*

Solution: It is obvious that the two tabs cannot be ignored. So, can they be used as part of the solution? If only we could somehow move the half rectangle to the left *up* in such a way that the bottom tab would then become part of the original rectangle, we would then have solved our puzzle. Let's do exactly that, cutting the rectangle in a zigzag fashion with each cut equal to the length of the tab:

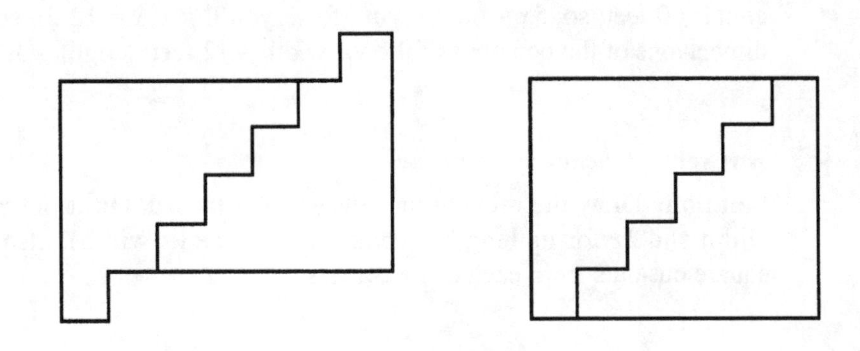

57 **Answer:** *With 6 cuts, 22 pieces of pie can be produced.*

Solution: Solving this puzzle by trial and error is cumbersome. So, consider simpler cases to see if a general pattern can be extracted from them.

Start by slicing the pie with one cut, which of course generates 2 pieces:

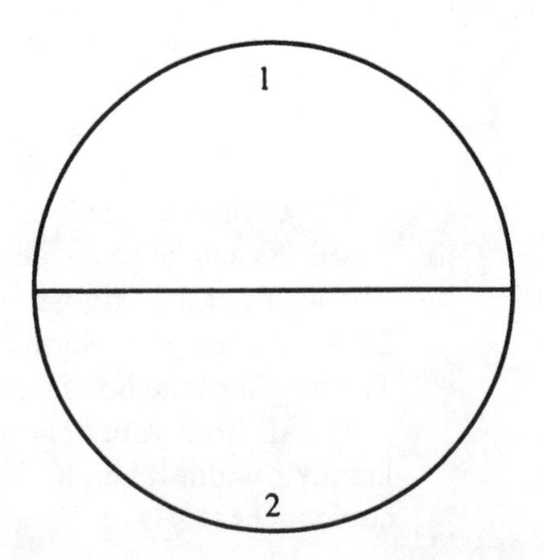

If you make a second cut, you will get 4 pieces:

A third cut generates 7 pieces:

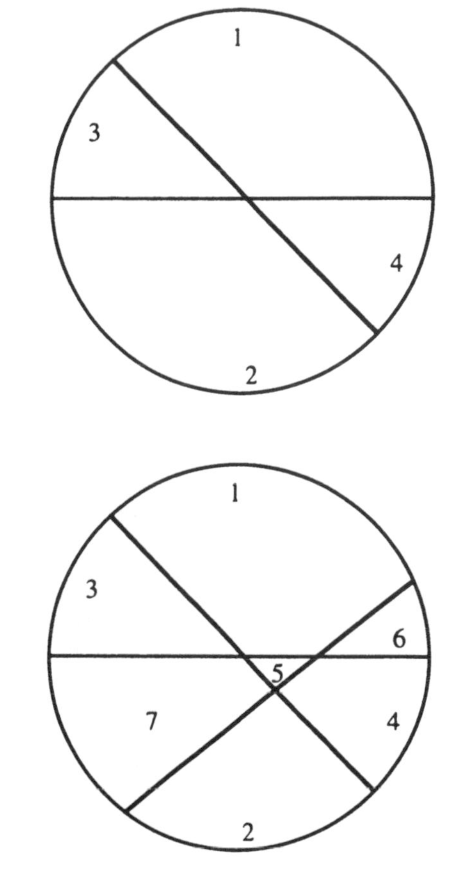

Tabulate the results so far:

Number of Cuts	Number of Pieces
1	2
2	4
3	7

Is there a pattern? Well, it seems that each new cut adds a number of pieces that is equal to the number of the cut. Concretely speaking, if the number of cuts is 1, then the number of new pieces is going to be 1. So, adding the total number of cuts to the total number of new pieces, we get 1 + 1 = 2 pieces. If the number of cuts is 2, then the number of new pieces is going to be 2. Adding this number to the previous number of pieces, we get 2 + 2 = 4. If the number of cuts is 3, then the number of new pieces is going to be 3. Add three new pieces of pie to the previous number of 4, we get 3 + 4 = 7 pieces.

This *algorithm*—or method of computing number of pieces from the number of cuts—predicts that a fourth cut would add 4 new pieces to the previous 7, thus generating 4 + 7 = 11 pieces. Does it indeed generate 11 pieces? Let's find out:

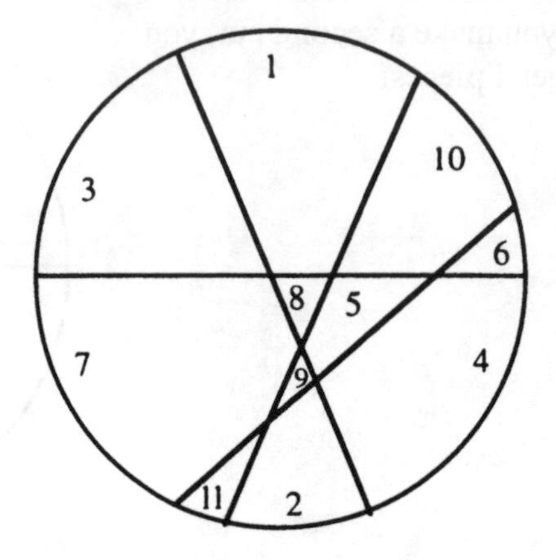

Yes it does, as predicted! So, now you can continue filling in your chart up to 6 cuts using the algorithm:

Number of Cuts	Number of Pieces
1	2
2	4
3	7
4	11
5	16
6	22

So, with 6 cuts, you can produce 22 pieces of pie.

8 Puzzles in Code Logic

Secret *codes* are of ancient origin. The sacred Jewish writers of ancient times, for instance, concealed their messages by reversing the alphabet, that is, by using the last letter of the alphabet in place of the first, the next last for the second, and so on. This system, called *atbash,* is exemplified in the Bible, in Jeremiah 25:26, where "Sheshech" is written for "Babel" (Babylon), using the second and twelfth letters from the end of the Hebrew alphabet instead of from the beginning. Spartan soldiers communicated with their field generals by means of messages written across the adjoining edges of a strip of parchment wrapped spirally around a staff called a *scytale.* Once unrolled, the message could be read only by wrapping the strip on an identical staff. In the Middle Ages, alchemists used astrological signs to send secret messages. By the time of the Renaissance, during the fourteenth and fifteenth centuries, treatises were written on techniques for deciphering such codes. Incidentally, one of the more famous code-makers of all time was the American writer Edgar Allan Poe (1809–1849).

How To...

The puzzles in this chapter are based on correspondences that have been set up between letters and numbers: for example, a = 1, b = 5, and so on. Your task will be to unravel the concealed correspondences.

Puzzle Properties

There are two main kinds of code puzzle: (1) the type in which numbers stand for letters of the alphabet, and (2) the type in which letters of the alphabet stand for numbers. Puzzles of the first type, known as *cryptograms,* presume that you know how words and sentences are constructed in ordinary English; puzzles of the second type, known as *alphametics,* presuppose that you know how simple arithmetical problems using digits are laid out and solved.

Example 1 Here's an example of how to go about solving a typical cryptogram puzzle.

Decipher the following message, given that 2 = S, 6 = P, 10 = B, and 12 = K. The message relates to something you are using right now.

1 0 4 2   4 2   5   6 3 7 7 8 9   10 11 11 12

The first thing to do is to replace the numbers 2, 6, 10, 12 with the given letters: 2 = S, 6 = P, 10 = B, 12 = K. This gives you an initial glimpse of the hidden message:

```
1  0  4  2     4  2     5     6  3  7  7  8  9     10  11  11  12
         ↑        ↑           ↑                    ↑           ↑
         S        S           P                    B           K
```

Note that 5 is a single-letter word. The only two possible candidates for single-letter words in English are the pronoun *I* and the indefinite article *a*. If you're not convinced of this, just go through the alphabet, and you will see that you cannot form a word with the other letters used singly: *b, c, d, e,* . . . . The pronoun *I* usually starts a sentence: *I am going out, I love puzzles,* and so on. There are, of course, sentences where this is not the case, but these are less frequent. Since the single-letter word 5 occurs within the sentence, it is best to start out with the most likely possibility, which in this case is the indefinite article *a*. So, replace 5 with A. You can always go back to this point in your solution if this replacement should not work out and replace 5 with I:

```
1  0  4  2     4  2     5     6  3  7  7  8  9     10  11  11  12
         ↑        ↑     ↑     ↑                    ↑           ↑
         S        S     A     P                    B           K
```

Now, focus your attention on the last word in the sentence: 10 11 11 12 = B ? ? K. Notice the sequence of two 11's. This means that the word contains two identical letters. A little reflection on how words are constructed should convince you that this sequence of double letters cannot be consonants, for that would mean that there are four consonants in a word without any vowel:

10 11 11 12 = B C C K

10 11 11 12 = B D D K

10 11 11 12 = B F F K

. . .

So, you can safely conclude that the sequence is made up of two identical vowels: *aa, ee, ii, oo,* or *uu.* The only replacement that produces a legitimate word is *oo.* The word is, therefore, 10 11 11 12 = B O O K. The message now looks like this:

1	0	4	2	4	2	5	6	3	7	7	8	9	10	11	11	12
			↑		↑	↑	↑						↑	↑	↑	↑
			S		S	A	P						B	O	O	K

Now, consider the two-letter word 4 2 = ? S. English word-structure forces you to conclude that only a vowel is possible as a replacement for 4. So the word could be *as, us,* or *is.* It is unlikely for either *as* or *us* to occur as the second word in this sentence. Try out a few sentences with *as* or *us* as the second word and you will quickly become convinced of this. So, once again, go with the most likely possibility, namely that 4 2 = I S. Once again, should this turn out to be erroneous, you can always go back to this point in your solution and choose one of your other two options (4 2 = A S, 4 2 = U S). Replace 4 with I in the first word as well:

1	0	4	2	4	2	5	6	3	7	7	8	9	10	11	11	12
		↑	↑	↑	↑	↑	↑						↑	↑	↑	↑
		I	S	I	S	A	P						B	O	O	K

Now, consider the first word. The likeliest candidate for a four-letter word in English starting a sentence and partially deciphered as ? ? I S is the demonstrative adjective *This*. So, 1 = T and 0 = H:

1	0	4	2	4	2	5	6	3	7	7	8	9	10	11	11	12
↑	↑	↑	↑	↑	↑	↑	↑						↑	↑	↑	↑
T	H	I	S	I	S	A	P						B	O	O	K

Finally, consider P 3 7 7 8 9. Use the other parts of the sentence to help you decipher this word: *This is a p . . . book.* You are looking for a kind of book that: (1) has six letters in it, (2) starts with the letter *p,* (3) has a sequence of two identical letters in it, and (4) is related to the given clue: *The message relates to something you are using right now.* Try out a few possibilities, and you will soon come to the conclusion that *puzzle* is the required word:

1	0	4	2	4	2	5	6	3	7	7	8	9	10	11	11	12
↑	↑	↑	↑	↑	↑	↑	↑	↑	↑	↑	↑	↑	↑	↑	↑	↑
T	H	I	S	I	S	A	P	U	Z	Z	L	E	B	O	O	K

As you can see, the solution is: *This is a puzzle book.*

Example 2 The term *alphametic* was first coined in 1955 by the American puzzlist J. A. H. Hunter to distinguish code puzzles where letters replace numerals in a random fashion from those in which the letters form meaningful words and phrases. *Alphametic* refers to the latter type. The originator of such puzzles was none other than the great English puzzlist Henry E. Dudeney (1847–1930), who referred to them as exercises in "verbal arithmetic."

The following is a classic in the genre. You are asked simply to figure out what numbers the letters represent.

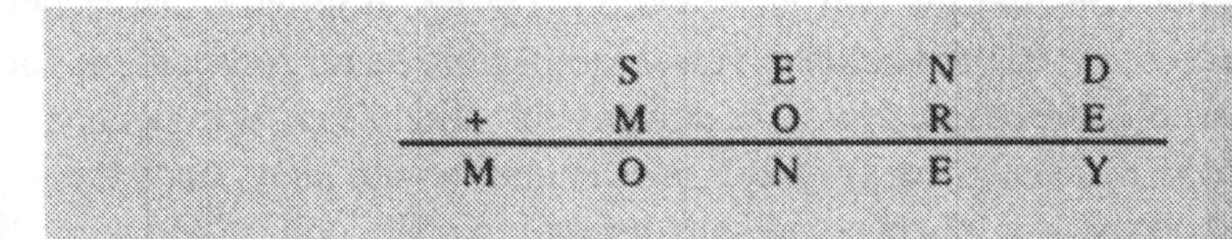

Since this is a coded problem in arithmetic, it can instantly be established that the **M** at the extreme left is a carryover digit equal to 1, because 1 is the only carryover possible when two digits are added together in the previous column—in this case **S** + **M**—even if the column has itself a carryover from the column before. If you do not see this, consider the column in question. There are 10 digits including zero. The maximum two different digits can add up to is 17; this would occur with the two largest digits, 9 and 8. So, let the two digits in the column equal 9 and 8, just for the sake of illustration, ignoring for the moment the actual letters that are there:

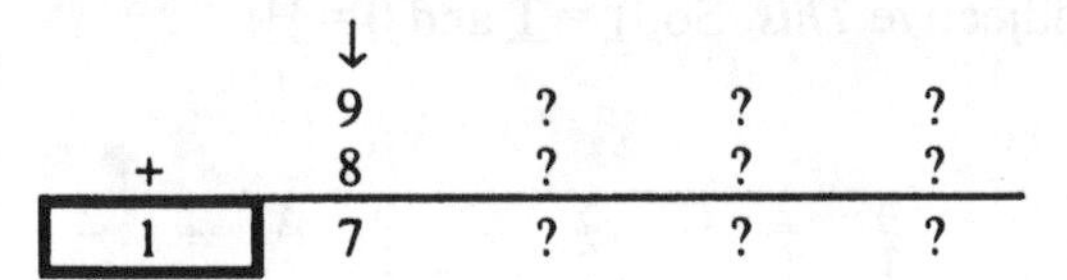

Now do you see that **M** can only equal 1? Even if there were a carryover from the previous column, the maximum 9 + 8 + 1 (carryover) could possibly add up to is 18. Having proved that **M** = 1, put its numerical value in the puzzle, noticing that it occurs in two places:

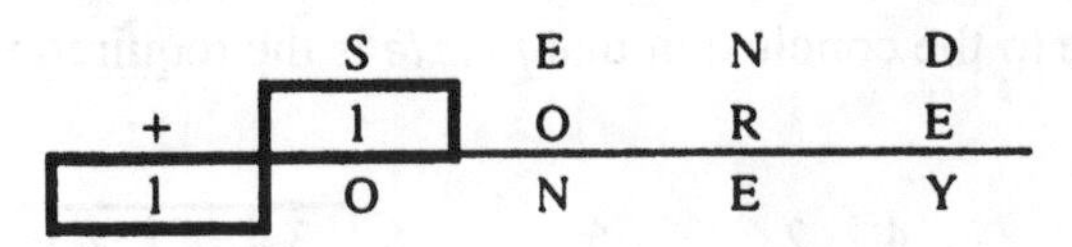

Now, focus on the **S** + 1 = **O** column. You have established that **M** = 1, so you know that **O** does not equal 1. It must therefore equal 0, since the **S** above it cannot be greater than 9. So, adding together 9 + 1 you would get 10. You would not get 11 because that would make the **O** below equal to 1, which, as you have just determined, is the value of **M**. So, go ahead and replace **O** = 0 in the places where it is found in the layout:

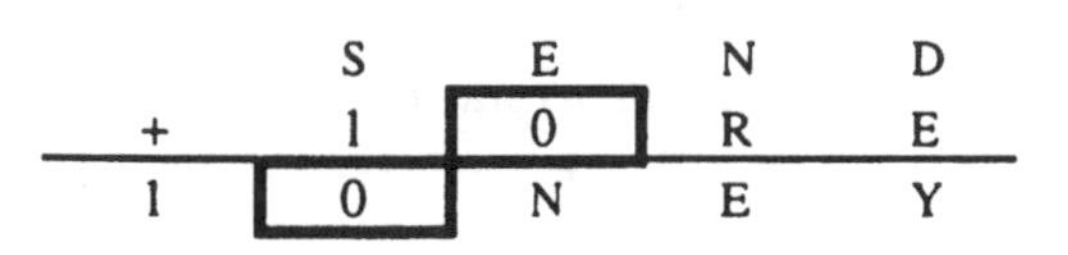

Now, consider the **S** in the **S** + 1 = 0 column. It must be either 9 or 8. The 8 takes into account a possible carryover from the center column. But, then, look at that column. It has a 0 in it: **E** + 0 = **N**. So, there is no way that this column can produce a carryover, even if **E** = 9, given that the digit 1 has been already assigned to **M**. So, there can be no carryover from the **E** + 0 = **N** column to the **S** + 1 = 0 column. You can thus safely conclude that **S** = 9:

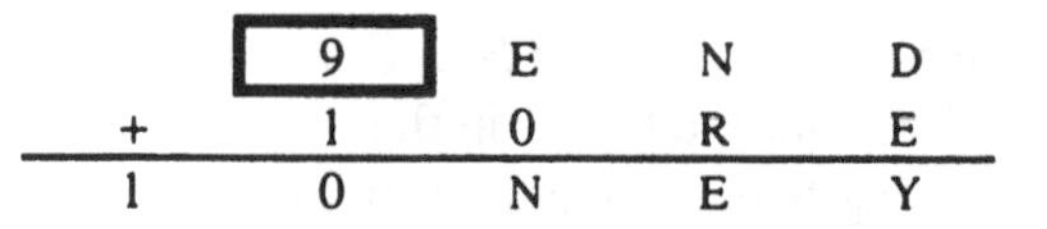

Now turn your attention to the **E** + 0 = **N** column. It must have a carryover from the previous **N** + **R** = **E** column, because if there were no carryover, this column would not make arithmetical sense. Why? Because any number added to 0 would produce that number as a sum: 2 + 0 = 2, 3 + 0 = 3, and so on. So, without a carryover, **E** + 0 should add up to **E**. But it doesn't. Therefore, **E** + 0 = **N** must have a carryover. You can show this arithmetically as 1 (carryover) + **E** + 0 = **N**.

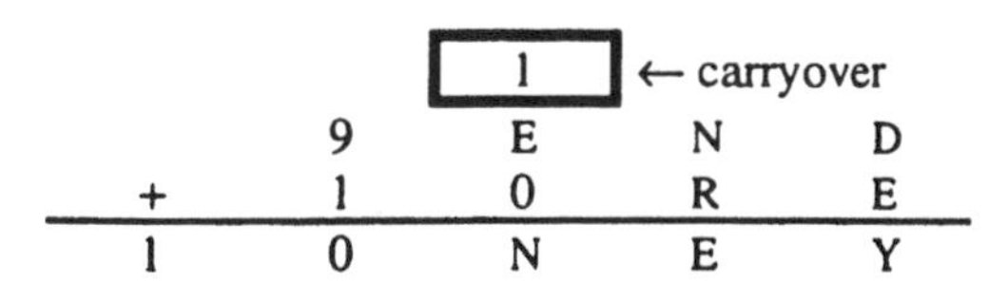

It follows that **N** = **E** + 1. Remember this fact! Now, consider the **N** + **R** = **E** column. As you have just determined, the addition of **N** + **R** produces a carryover. In arithmetical terms, this means that the sum of **N** + **R**, which is given to you as **E**, is greater than 10. This fact can be represented, of course, as **E** + 10. So, now you know that **N** + **R** = **E** + 10. This assumes, however, that there is no carryover from the previous column. If there is a carryover from that column, then the appropriate equation is 1 (carryover) + **N** + **R** = **E** + 10, which in reduced form is **N** + **R** = **E** + 9.

You have established that **N** = **E** + 1. So, substitute this value in the two possibilities for the column:

(1) *With no carryover:*

N + **R** = **E** + 10 → **E** + 1 + **R** = **E** + 10 → **R** = 9

(2) *With a carryover:*

N + **R** = **E** + 9 → **E** + 1 + **R** = **E** + 9 → **R** = 8

Therefore, **R** = 9 or 8. You have already established that **S** = 9. So, you can conclude that **R** = 8:

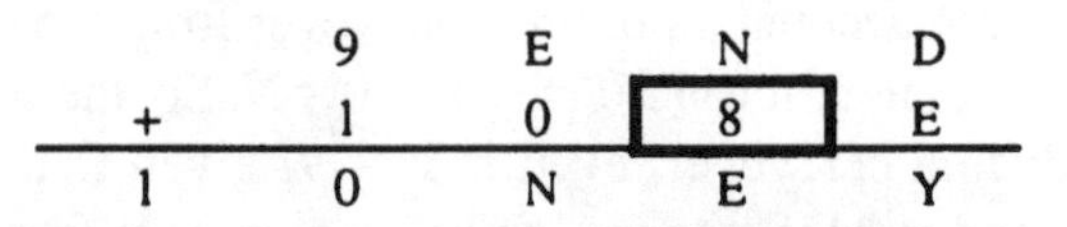

Consider the **N** + 8 = **E** column again. The digits 0 and 1 have already been assigned, so you can deduce that this column produces a carryover, since **N** is at minimum 2 (not 0 or 1). Now, it is obvious that the **D** + **E** = **Y** column will also produce a carryover, since **N** or **E** cannot be 0 or 1. **N** also cannot be 2, for then **E** would be equal to 0 or 1 in the **N** + 8 = **E** column, which is impossible since these two digits have already been assigned. Try out a few possibilities, starting with **N** = 3. Indicate all your hypothetical replacements in a different color. (We have used italic type.)

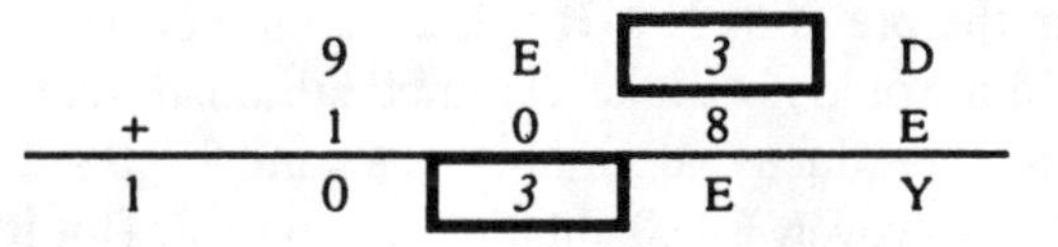

From the **E** + 0 = 3 column, you can see that **E** = 2, given that the column has a carryover. Put this value in the appropriate locations:

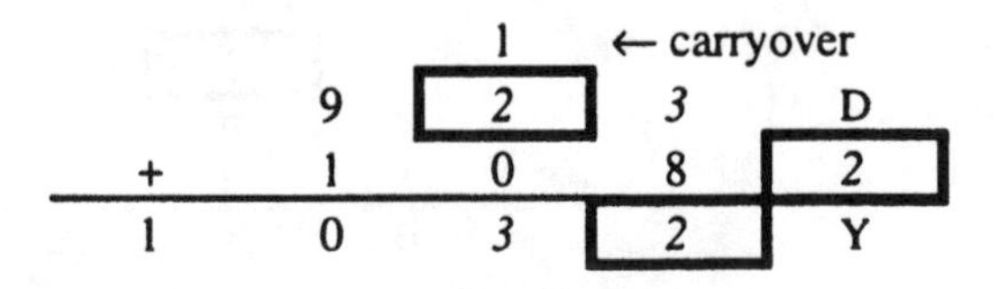

The 3 + 8 = 2 column implies a carryover from the previous column. So, the **D** in this column must be 8 or 9. But this is impossible, since these digits have been already assigned. So, discard the original hypothesis, namely that **N** = 3. Try **N** = 4 instead:

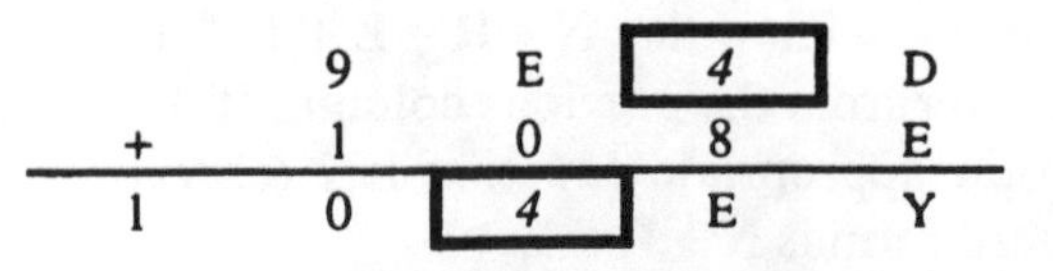

This replacement makes **E** = 3:

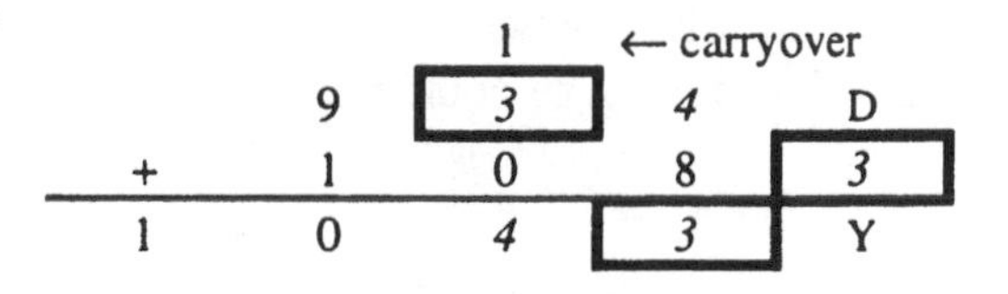

Now, **D** must be equal to 7, 8, or 9. You have already assigned 8 and 9, so these two possibilities can be discarded. You can also discard 7, because if **D** = 7, then **Y** = 0. But 0 has already been assigned. Once again, discard the working hypothesis that **N** = 4, and go on to **N** = 5 as your next hypothesis:

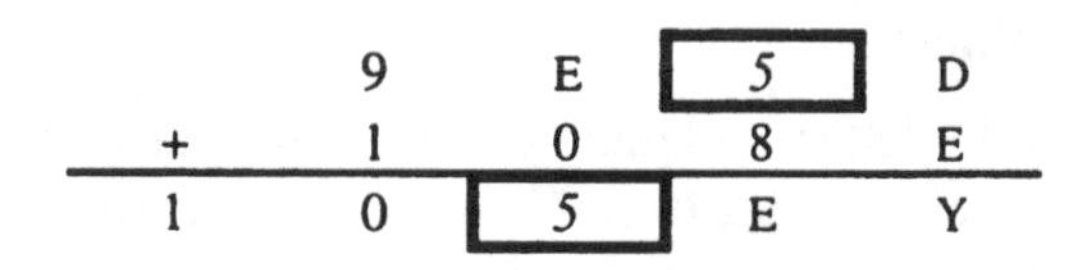

This replacement makes **E** = 4:

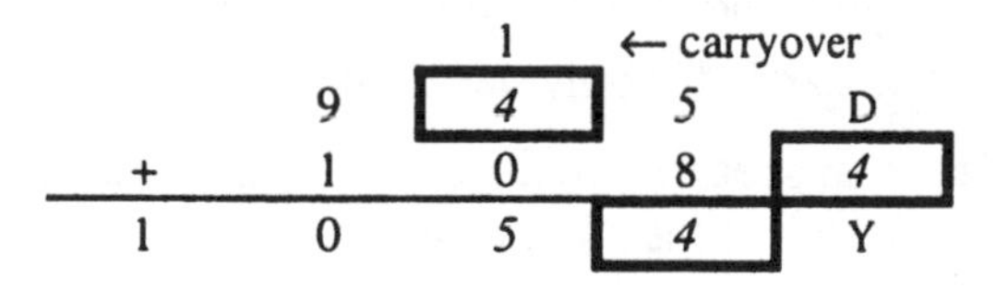

Now, it can be seen that **D** = 6, 7, 8, or 9. You can discard 8 and 9, since these digits have already been assigned. You can also discard **D** = 7, because this would make **Y** = 1 in the column, and this digit too has been assigned. And you can reject **D** = 6, because this would make **Y** = 0 in the column, which has also been assigned. So, reject the hypothesis that **N** = 5, and go on to **N** = 6 as your next hypothesis:

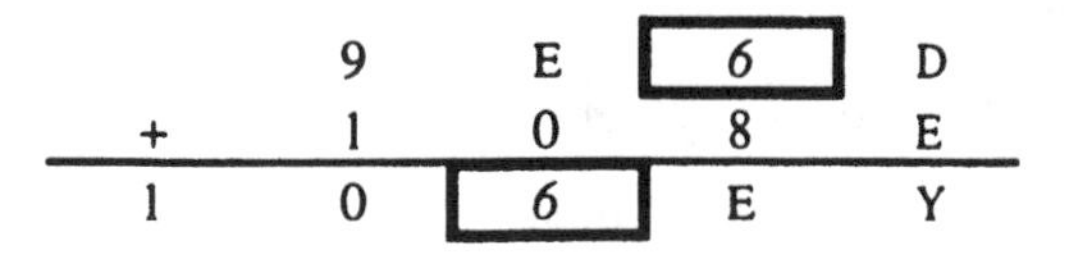

This replacement makes **E** = 5:

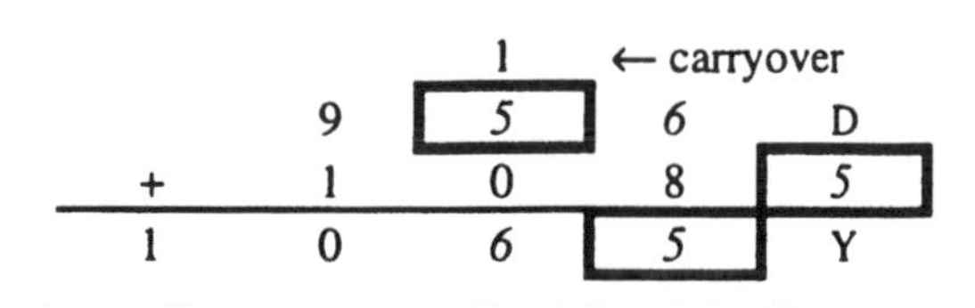

This means that **D** = 6, 7, 8, or 9. As before, discard 8 and 9, because these have already been assigned. Discard **D** = 6 as well, because, that would make **Y** = 1; but 1 has already been assigned. However **D** = 7 works perfectly, since 7 + 5 = 12. This means, of course, that **Y** = 2. The solution is now complete:

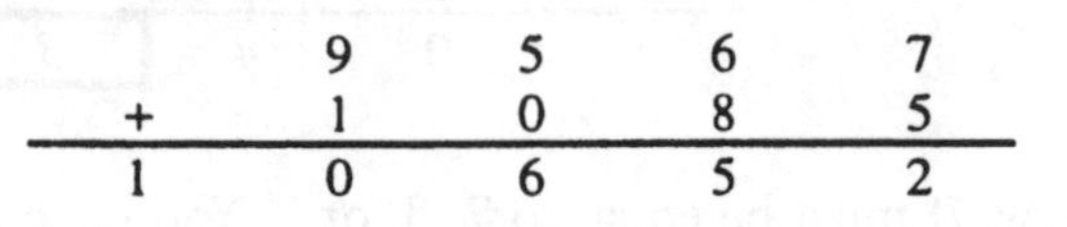

	9	5	6	7
+	1	0	8	5
1	0	6	5	2

Summary

In summary, when solving cryptograms (puzzles 58–61), keep in mind the following things:

- Look for letters or letter combinations that would seem to make sense in certain positions: for example, a three-letter word starting a sentence is highly likely to be the definite article *the.*
- Always keep the possible meaning of the message in mind when deciphering certain words.

When solving alphametics (puzzles 62–66), keep in mind these points:

- A letter stands for only one digit.
- Letters in specific positions can represent only certain digits (recall from example 2 that the **M** at the extreme left could only be equal to 1).
- Letters in specific positions cannot stand for certain numbers (in example 2, **M** could not possibly have been equal to 0 because that digit can never occur at the extreme left of an addition problem).

To help you get started on alphametics, note the following properties of the numbers "1" and "0":

- *Properties of "1"*

  1 multiplied by itself = 1

  1 divided by itself = 1

  1 multiplied by any number = that number

  1 divided into any number = that number

- *Properties of "0"*

  0 multiplied by itself = 0

  0 multiplied by any number = 0

  0 added to itself = 0

  0 added to any number = that number

  0 subtracted from any number = that number

□□■ Puzzles 58–66

Answers, along with step-by-step solutions, can be found at the end of the chapter.

58 If 2 = L, 5 = T, and 7 = S, can you decipher the following message, which concerns a certain sport or physical activity that someone likes?

1   2 1 3 4   5 6   7 8 1 9   1 10   7 11 9 9 4 12

59 Decipher the following message, which is something normally uttered on the phone. Incidentally, 1 = H:

1 2 3 3 4 , 5 1 4   6 7   6 8 ?

60 By decoding the following message, you will get a colloquial expression for taking leave of someone. Note that, 2 = L:

1 ' 2 2   3 4 4   5 6 7   2 8 9 4 10

61 This message will tell you something that by now you have come surely to know, after solving so many puzzles. Incidentally, 1 = L, 9 = E, and 13 = Y:

1 2 3 4 5   6 7 8 8 1 9 10   11 12 9   9 11 10 13

62 Here's an addition problem that seems to add up to an interrogative: *Two* + *Two* = *Who?* Can you decode it numerically?

```
  T  W  O
+ T  W  O
---------
  W  H  O
```

63 This is an interesting multiplication alphametic in which *it* is multiplied by *it*self! Find the numbers that the letters represent.

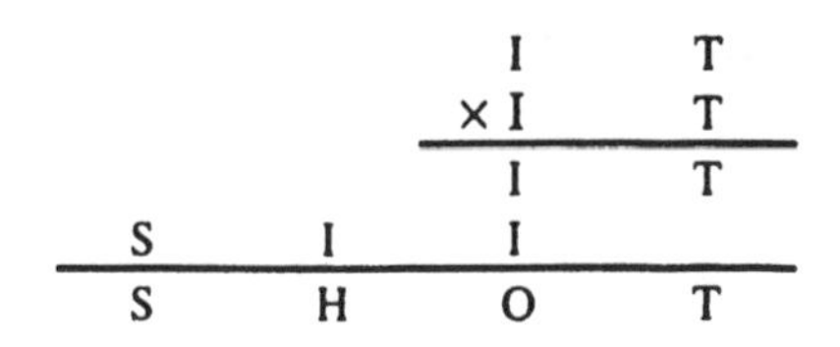

```
         I  T
       × I  T
       ------
         I  T
   S  I  I
   ----------
   S  H  O  T
```

64 *Too go* or not *too go*? What does it all mean—numerically of course?

```
  T  O  O
−    G  O
---------
     G  O
```

65 Does *To* × *To* really equal *TNN*? You'll know for sure if you can find the numbers that the letters represent.

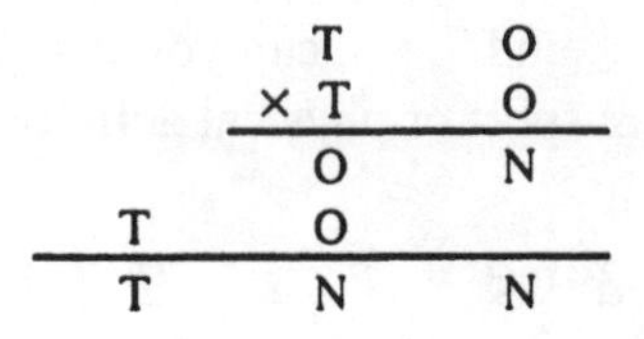

This is a message that a brother left his sister: *Trot* + *trot* = *room*. He wanted her, of course, to trot right on to her room. But what does it mean numerically?

	T	R	O	T
+	T	R	O	T
	R	O	O	M

Answers and Solutions

58 **Answer:** *I like to swim in summer.*

Solution: The first thing to do is to replace the numbers with the given letters: namely, 2 = L, 5 = T, 7 = S:

1	2	1	3	4	5	6	7	8	1	9	1	10	7	11	9	9	4	12
	↑ L				↑ T		↑ S						↑ S					

The easiest word to figure out is 5 6, since you can see that it starts with T. The word can only be *to*:

1	2	1	3	4	5	6	7	8	1	9	1	10	7	11	9	9	4	12
	↑ L				↑ T	↑ O	↑ S						↑ S					

Now, consider 1 = ?, a single-letter word starting off a sentence. It can only be, therefore, either the indefinite article *A* or the subject pronoun *I*. If you choose *A*, then you get:

1	2	1	3	4	5	6	7	8	1	9	1	10	7	11	9	9	4	12
↑ A	↑ L	A			↑ T	↑ O	↑ S		↑ A		↑ A		↑ S					

Now, the word 1 10 = A ? can only be *an*, because *at* would be A T = 1 5 and *as* would be A S = 1 7:

1	2	1	3	4	5	6	7	8	1	9	1	10	7	11	9	9	4	12
↑	↑	↑			↑	↑	↑		↑		↑		↑					
A	L	A			T	O	S		A		A	N	S					

But this is impossible, because the following word starts with a consonant—it is grammatically incorrect to use the form *an* before a noun or adjective starting with a consonant. So, reject 1 = A, concluding instead that 1 = I:

1	2	1	3	4	5	6	7	8	1	9	1	10	7	11	9	9	4	12
↑	↑	↑			↑	↑	↑		↑		↑		↑					
I	L	I			T	O	S		I		I		S					

Now, consider the letter represented by 9. You can see that it occurs in S 8 I 9 and as a double letter in S 11 9 9 4 12. This means that it can only be a consonant, for a vowel could not occur in both those positions. So, 4 is a vowel, since only a vowel can fit meaningfully into the space after the double consonant in S 11 9 9 4 12 and into the space at the end of L I 3 4. The best bet for 4, therefore, seems to be E:

1	2	1	3	4	5	6	7	8	1	9	1	10	7	11	9	9	4	12
↑	↑	↑		↑	↑	↑	↑		↑		↑		↑				↑	
I	L	I		E	T	O	S		I		I		S				E	

There are three possibilities for the first two words: *I live, I like, I line.* In the context of the clue given in the puzzle, namely that the message concerns a sport or physical activity that someone *likes,* the most likely possibility is *I like.* So, let 3 = K:

1	2	1	3	4	5	6	7	8	1	9	1	10	7	11	9	9	4	12
↑	↑	↑	↑	↑	↑	↑	↑		↑		↑		↑				↑	
I	L	I	K	E	T	O	S		I		I		S				E	

The message that is emerging is: *I like to do something*—a sport or physical activity. A little reflection, should easily convince you that the word S 8 I 9 is *swim.* This means that 8 = W and that 9 = M:

1	2	1	3	4	5	6	7	8	1	9	1	10	7	11	9	9	4	12
↑	↑	↑	↑	↑	↑	↑	↑	↑	↑	↑	↑		↑		↑	↑	↑	
I	L	I	K	E	T	O	S	W	I	M	I		S		M	M	E	

Look at the message now, and it should take you very little to figure out what the rest of it is: *I like to swim in summer.*

1	2	1	3	4	5	6	7	8	1	9	1	10	7	11	9	9	4	12
↑	↑	↑	↑	↑	↑	↑	↑	↑	↑	↑	↑	↑	↑	↑	↑	↑	↑	↑
I	L	I	K	E	T	O	S	W	I	M	I	N	S	U	M	M	E	R

59 **Answer:** *Hello, who is it?*

Solution: In this case, note that: (1) the message concerns something that is uttered over the phone, (2) a comma occurs after the first word, (3) it is a question (since it ends with a question mark), and (4) that 1 = H:

1	2	3	3	4	,	5	1	4	6	7	6	8	?
↑					↑		↑						↑
H					,		H						?

Given the above facts, the only possibility for the first word is *Hello.* So, 2 = E, 3 = L, 4 = O:

1	2	3	3	4	,	5	1	4	6	7	6	8	?
↑	↑	↑	↑	↑	↑		↑	↑					↑
H	E	L	L	O	,		H	O					?

And the only possibility for the word 5 H O is, clearly, *who.* So, 5 = W:

1	2	3	3	4	,	5	1	4	6	7	6	8	?
↑	↑	↑	↑	↑	↑	↑	↑	↑					↑
H	E	L	L	O	,	W	H	O					?

Now, consider the last two words in the context of what the message is turning out to be: *Hello, who . . . ?* A little reflection should convince you that the last two words are probably *is* and *it*, making 6 = I. And, in fact, this replacement allows you to complete the message logically:

1	2	3	3	4	,	5	1	4	6	7	6	8	?
↑	↑	↑	↑	↑	↑	↑	↑	↑	↑	↑	↑	↑	↑
H	E	L	L	O	,	W	H	O	I	S	I	T	?

60 **Answer:** *I'll see you later.*

Solution: Register the given fact that 2 = L, and note that there is an apostrophe in the first word:

1	'	2	2	3	4	4	5	6	7	2	8	9	4	10
	↑	↑	↑							↑				
	'	L	L							L				

The first word is obviously a contraction. Thus, the only possibility for this word is *I'll.* So, 1 = I:

1	'	2	2	3	4	4	5	6	7	2	8	9	4	10
↑	↑	↑	↑							↑				
I	'	L	L							L				

The second word has a double letter in it: 3 4 4 = ? ? ?. Assume that 4 4 is a double consonant. Then, the first letter in the word is necessarily a vowel. If 3 = A, then the only possible word for 3 4 4 is *add—all* is not a possibility because A L L would be equal to 3 2 2, not 3 4 4. But *add* does not seem to lead anywhere, given that the message is a colloquial expression for taking leave of someone (i.e., for saying good-bye). Of the remaining vowel possibilities for 3—namely, E, I, O, U—it can also be seen that no meaningful word can be made for 3 4 4. So, 3 is not a vowel, and therefore 4 4 is not a double consonant. Hence, the 4 4 in 3 4 4 is a double vowel. A little reflection should convince you that the word is *see.* If you are not convinced, try to fit other words into the sequence *I'll* 3 4 4. You'll soon find that no other word works out. So, 3 = S and 4 = E:

1	'	2	2	3	4	4	5	6	7	2	8	9	4	10
↑	↑	↑	↑	↑	↑	↑				↑			↑	
I	'	L	L	S	E	E				L			E	

Now, you can complete the solution easily. Consider what you have decoded so far: *I'll see . . . l. . . .* You are looking for a colloquial way of saying good-bye. The only possibility is *I'll see you later:*

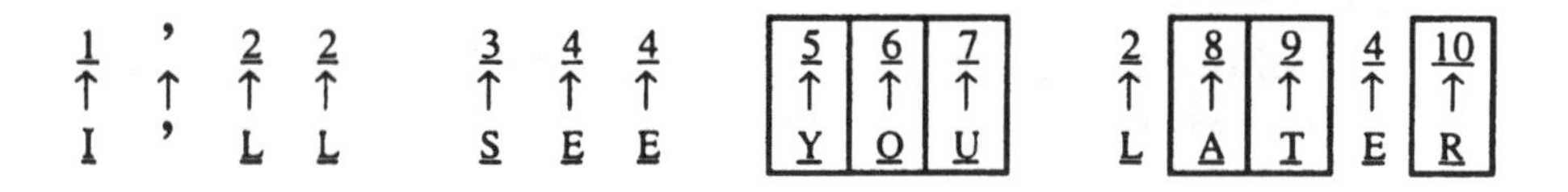

61 **Answer:** *Logic puzzles are easy.*

Solution: Register the fact that 1 = L, 9 = E, and 13 = Y:

1	2	3	4	5	6	7	8	8	1	9	10	11	12	9	9	11	10	13
↑									↑	↑				↑	↑			↑
L									L	E				E	E			Y

A little reflection should make it obvious that there is only one word that E ? ? Y can be—namely, **easy.** So, 11 = A and 10 = S:

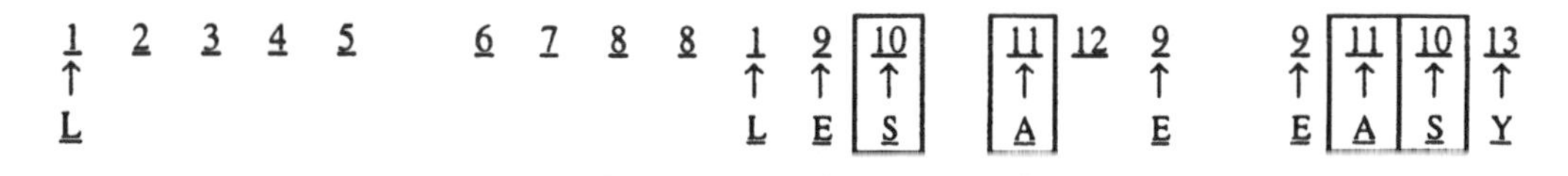

You can now infer that the second-last word A ? E is *are*. So, 12 = R:

1	2	3	4	5	6	7	8	8	1	9	10	11	12	9	9	11	10	13
↑									↑	↑	↑	↑	↑	↑	↑	↑	↑	↑
L									L	E	S	A	R	E	E	A	S	Y

Now, the second word 6 7 8 8 L E S has a double letter in it. In the context of the given clue—that the message refers to something that you have come to know, after working through so many puzzles—the most likely candidate for this word is *puzzles*. Let's work on this assumption. If it turns out to be an erroneous one, then we can always go back to this point, and try something else. So, 6 = P, 7 = U, and 8 = Z:

1	2	3	4	5	6	7	8	8	1	9	10	11	12	9	9	11	10	13
↑					↑	↑	↑	↑	↑	↑	↑	↑	↑	↑	↑	↑	↑	↑
L					P	U	Z	Z	L	E	S	A	R	E	E	A	S	Y

Finally, consider the first word: *L. . . puzzles are easy.* In the context of the given clue, and considering that the first word has five different letters in it, the only word that makes sense is *logic.*

1	2	3	4	5	6	7	8	8	1	9	10	11	12	9	9	11	10	13
↑	↑	↑	↑	↑	↑	↑	↑	↑	↑	↑	↑	↑	↑	↑	↑	↑	↑	↑
L	O	G	I	C	P	U	Z	Z	L	E	S	A	R	E	E	A	S	Y

62 **Answer:** *120 + 120 = 240 or 240 + 240 = 480.*

Solution: Right away, you can deduce that **O** = 0, since the only addition that is possible in the **O** + **O** = **O** column is 0 + 0 = 0.

T	W	0
+ T	W	0
W	H	0

So, there is no carryover to the middle column. And there is no carryover from the **T** + **T** = **W** column either, as the layout shows. So, **T** is less than 5.

Try out **T** = 4, using a different color to indicate that this is only a working hypothesis:

4	W	0
+ *4*	W	0
8	H	0

This means that **W** = 8. But if it does indeed equal 8, then the middle column will produce a carryover: **W** + **W** = 8 + 8 = 16. This would then make the **W** in the leftmost column equal to 9—that is, 1 (carryover) + 4 + 4 = 9. This is obviously a contradiction. So, discard the hypothesis that **T** = 4.

Try out **T** = 3 instead:

3	W	0
+ *3*	W	0
6	H	0

This means that **W** = 6. But if it does indeed equal 6, then the middle column will produce a carryover: **W** + **W** = 6 + 6 = 12. This would then make the **W** in the leftmost column equal to 7—that is, 1 (carryover) + 3 + 3 = 7. This is again a contradiction. So, discard the hypothesis that **T** = 3.

Try **T** = 2 instead:

2	W	0
+ 2	W	0
4	H	0

This means that **W** = 4. Then, in the middle column: **W** + **W** = 4 + 4 = 8. This is indeed possible. And, in fact, the puzzle is now solved.

2	4	0
+ 2	4	0
4	8	0

Let's see if it is a unique solution by letting **T** = 1:

1	W	0
+ *1*	W	0
2	H	0

This means that **W** = 2. Then, in the middle column: **W** + **W** = 2 + 2 = 4. This, too, is possible:

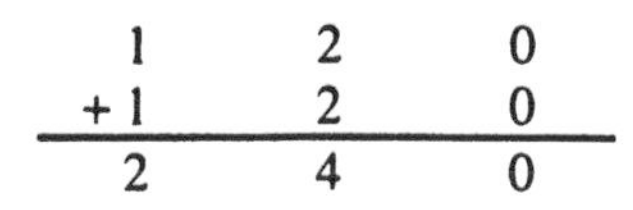

1	2	0
+ 1	2	0
2	4	0

In sum, two solutions to this alphametic are possible: 120 + 120 = 240 or 240 + 240 = 480. Note that the second solution is really twice the first: 240 + 240 = 480 = 2(120) + 2(120) = 2(240).

63 **Answer:** *$61 \times 61 = 3721$ or $51 \times 51 = 2601$.*

Solution: Note that multiplying a number by **T** produces that number as the product: that is, $\mathbf{T} \times \mathbf{T} = \mathbf{T}$ and $\mathbf{T} \times \mathbf{I} = \mathbf{I}$. So, you can safely conclude that $\mathbf{T} = 1$:

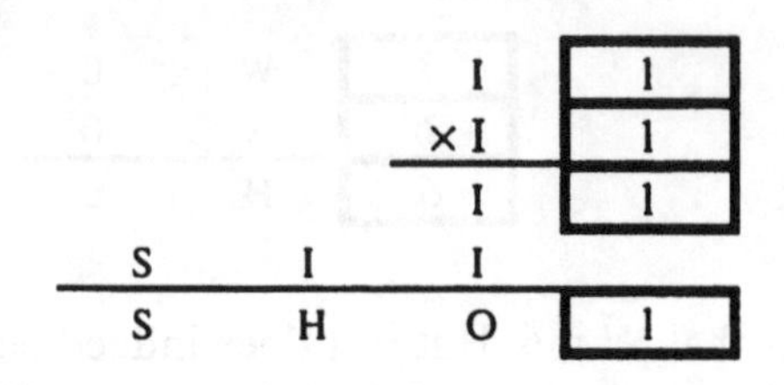

Next, note that there must be a carryover, because **I** and **H** cannot be the same.

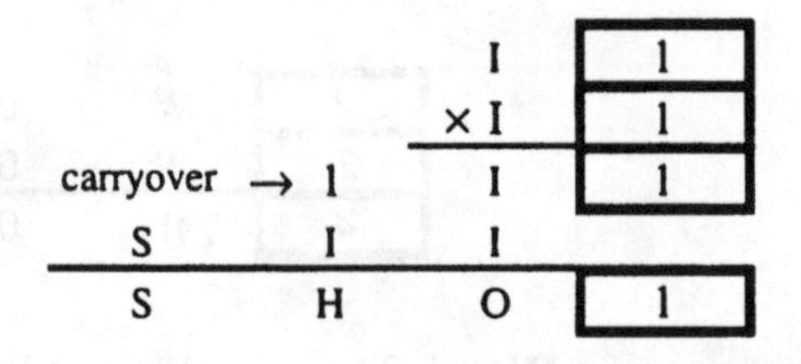

This carryover results, of course, from adding $\mathbf{I} + \mathbf{I}$ in the previous column. Therefore, the **I** is 5 or greater.

Try $\mathbf{I} = 9$, using a different color to indicate that this is only a hypothesis:

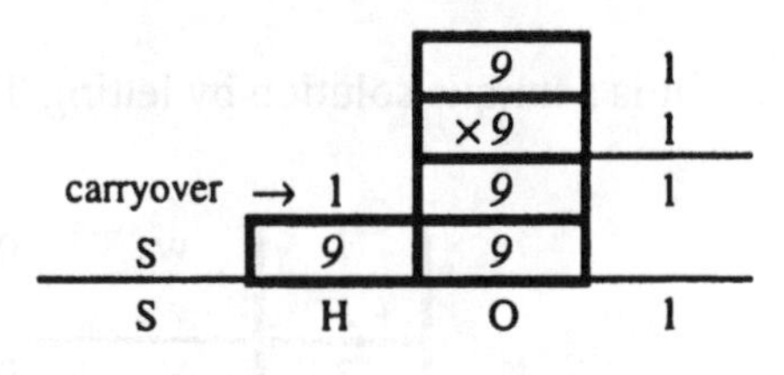

This doesn't work, because when you actually multiply 9×9, you get 81 as a product. But, as you can see, this is not a number ending in 9, as required by the substitution $\mathbf{I} = 9$.

Try $\mathbf{I} = 8$ instead:

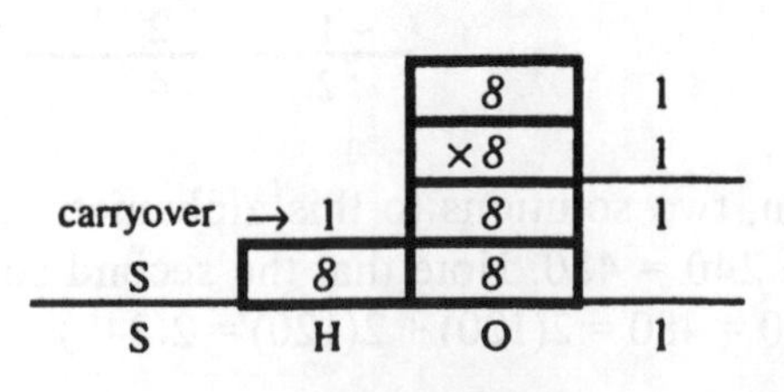

Once again, this produces an erroneous result, because when you actually multiply 8 × 8, you get 64 as a product, which is not a number ending in 8, as required by the substitution **I** = 8.

Try **I** = 7:

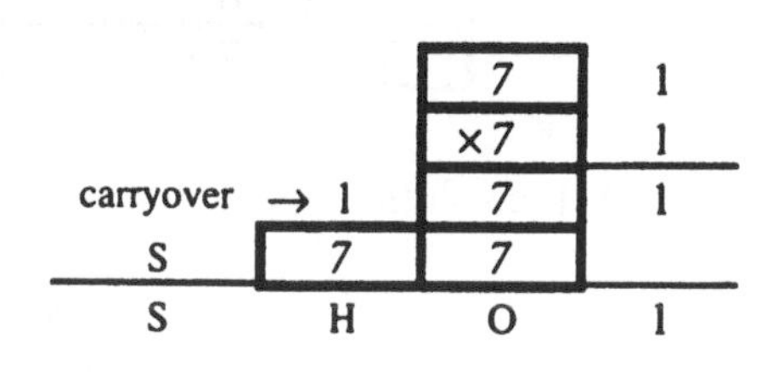

Again, this produces a contradictory result, because when you actually multiply 7 × 7, you get 49, which is not a number ending in 7, as required by the substitution **I** = 7.

Try **I** = 6:

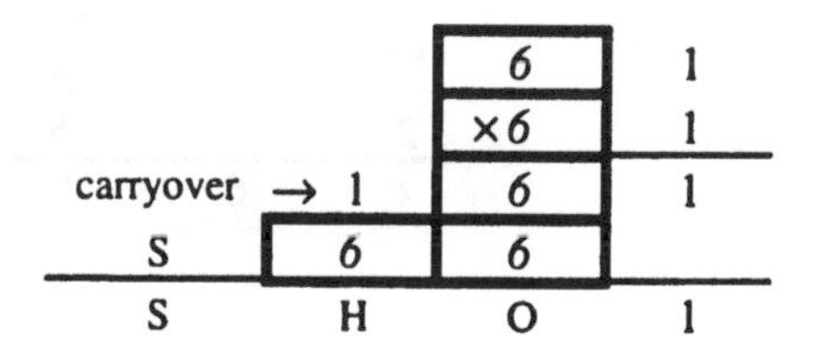

This produces a correct result, because when you actually multiply 6 × 6, you get 36, which is a number ending in 6, as required by the substitution **I** = 6. This means that **O** = 2, since 6 + 6 = 12:

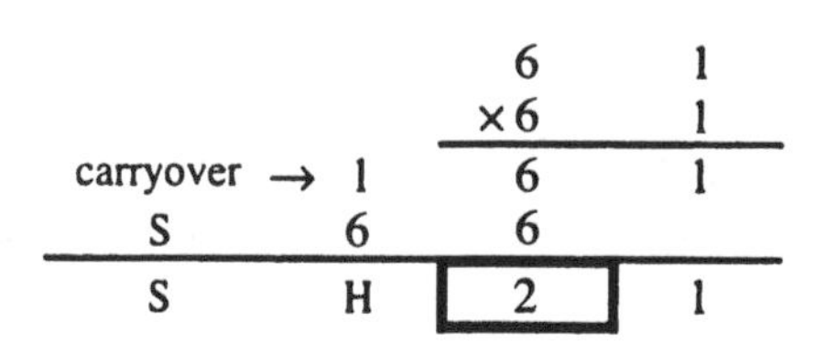

You can now see that **H** = 7, since the addition of 1 (carryover) + 6 = 7:

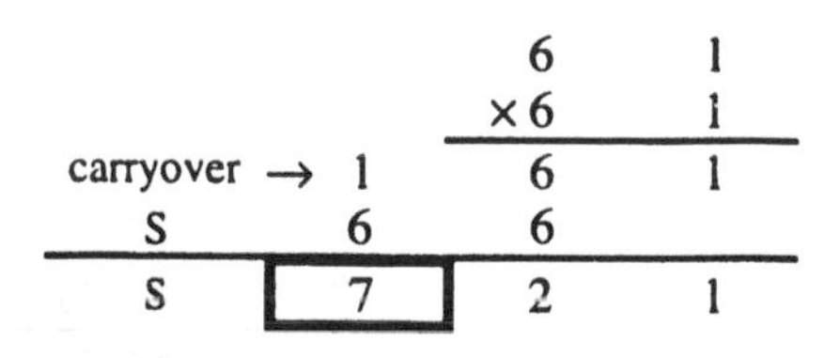

Finally, since $6 \times 6 = 36$, then **S** = 3.

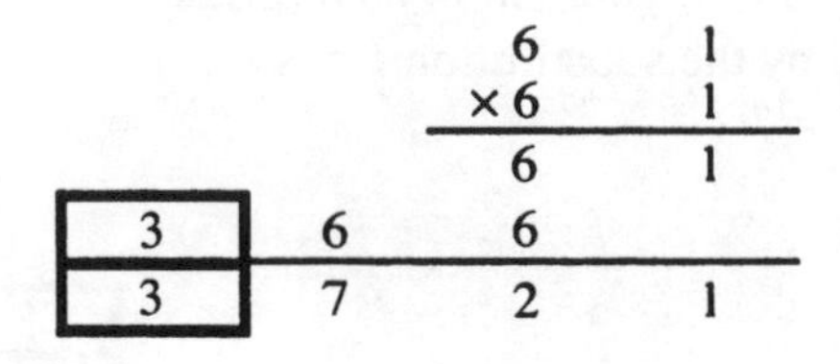

This completes the solution:

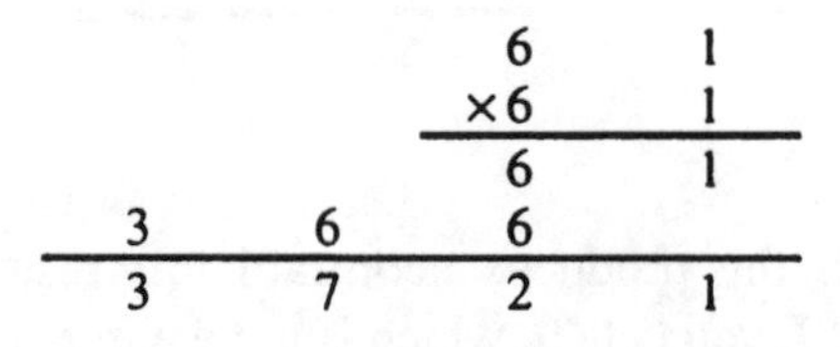

Incidentally, **I** = 5 leads to a second possible solution:

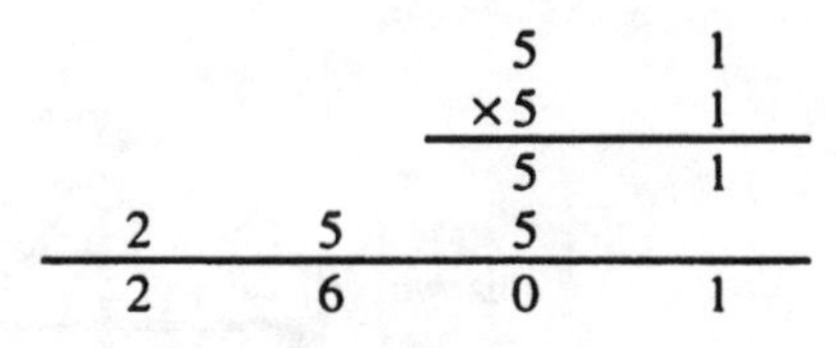

64 **Answer:** *100 – 50 = 50.*

Solution: Obviously, **O** = 0, given that 0 is the only digit that will work out arithmetically in the right column: **O – O = O**:

T	0	0
–	G	0
	G	0

Now, it can easily be seen that **T** = 1. If it were any larger, there would have to be a number below it underneath the subtraction line:

1	0	0
–	G	0
	G	0

Now, it is easy to see that **G** = 5, for that is the only number that works out arithmetically in the middle column: **O – G = G**:

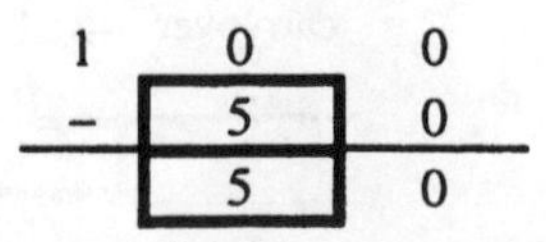

65 **Answer:** *12 × 12 = 144.*

Solution: Note that **T** times a number gives that number as a product. This means, of course, that **T** = 1:

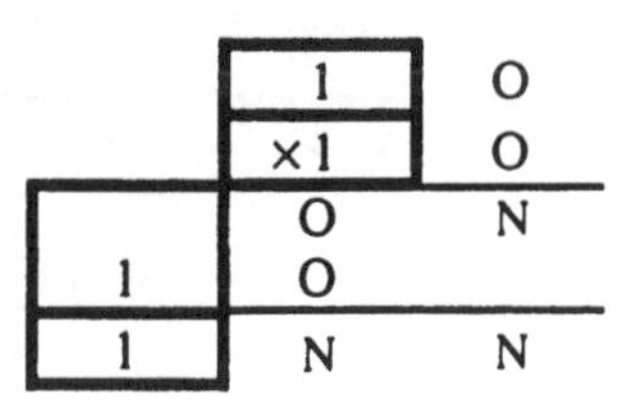

Obviously, **O** cannot be greater than 3, otherwise when you multiply **O** × **O**, you would get a carryover, contrary to what the puzzle layout shows. So, it is either 2 or 3. Keep in mind that **O** added to itself or multiplied by itself produces the same result, namely **N**. Try **O** = 3 first:

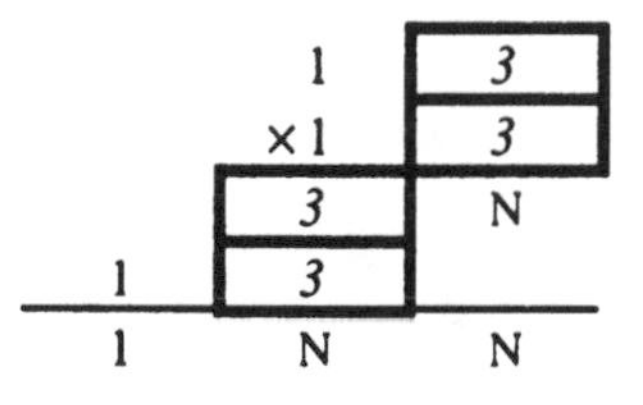

This means, of course, that 3 × 3 = 9. So, **N** = 9. But when you add 3 + 3 in the middle column, you get 6, not 9. So, the substitution **O** = 3 leads to an impossibility. The remaining choice, **O** = 2, fits the requirements: 2 × 2 = 4 and 2 + 2 = 4. This means, of course, that **N** = 4:

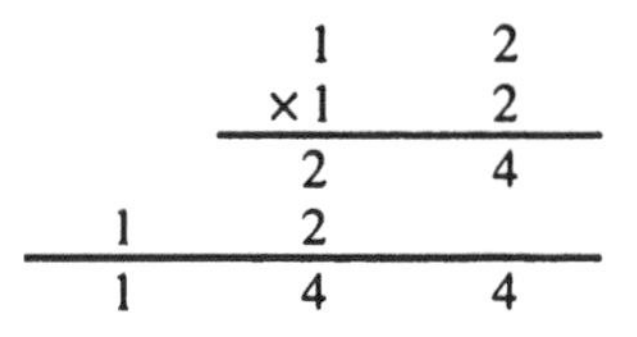

66 **Answer:** *2502 + 2502 = 5004.*

Solution: Note that **O** = 0, because it is the only number that would work in the **O** + **O** = **O** column:

T R 0 T
+T R 0 T
R 0 0 M

It is obvious now that **R** + **R** = 10, and, therefore, that **R** = 5:

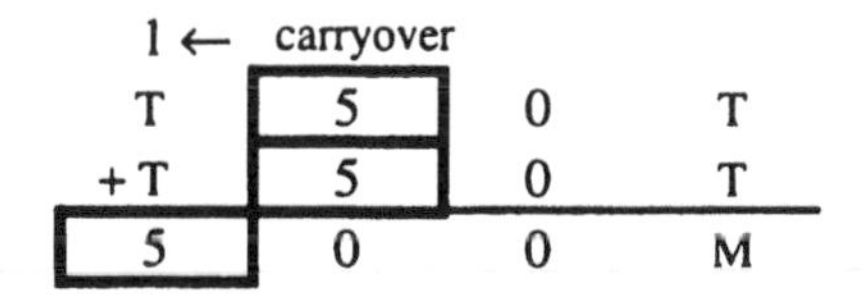

This means that **T** = 2, since 1 (carryover) + **T** + **T** = 5:

1 ←	carryover		
2	5	0	2
+2	5	0	2
5	0	0	M

Finally, it can be easily seen that **M** = 4:

2	5	0	2
+2	5	0	2
5	0	0	4

9 Puzzles in Time Logic

Time is truly a puzzle. From the paradoxes of Zeno of Elea (c. 495–435 B.C.) in ancient Greece to Albert Einstein's (1879–1955) theory of relativity in the twentieth century, time has always been the target of much fascination and debate by scientists and philosophers. Time has also been the source of many intriguing puzzles throughout the history of mathematics.

How To...

Like other kinds of puzzles, most puzzles in time logic can best be approached by setting up an appropriate diagram or chart that will help you visualize, or keep track of, the time arrangements, relations, or sequences involved.

Puzzle Properties

The two most frequent types of time puzzles are: (1) the type that requires you to figure out when something occurs on a day of the week, in a month of the year, and so on (example 1 below), and (2) the type that involves understanding the "mechanics" of a clock—that is, how a clock registers the passage of time (example 2 below). The latter is sometimes called a "clock puzzle."

Example 1 The following illustrates a time puzzle of the first type.

> Al, Bill, George, Don, and Ed have their birthdays on consecutive days, but not necessarily in that order.
>
> 1. Al's birthday is as many days before George's as Bill's is after Ed's.
> 2. Don is two days older than Ed.
> 3. George's birthday is on Wednesday.

A simple *weekday chart,* such as the following one, will allow you to keep track of the various possibilities or scenarios as you go along:

MONDAY	TUESDAY	WEDNESDAY	THURSDAY	FRIDAY	SATURDAY	SUNDAY

Clue 3 provides a definite point of reference to insert into the chart, since it tells you that George's birthday is on Wednesday. So, start by putting George into the Wednesday cell:

MONDAY	TUESDAY	WEDNESDAY	THURSDAY	FRIDAY	SATURDAY	SUNDAY
		GEORGE				

Clue 2 tells you that Don is two days older than Ed. This means that his birthday falls two days before Ed's birthday. If you do not see this, consider a concrete example. Let's say that Don's birthday falls on Monday, October 1, and that Ed's birthday falls on Wednesday, October 3. Both were born in the same year. Who is older? Don is. By how many days? By two days, of course. To put it another way, *Don is two days older than Ed.*

Clue 2 therefore suggests several hypothetical scenarios. You can exclude two scenarios from the outset: (1) putting Don in a Monday cell, since this would assign Ed to a Wednesday cell (but that cell is already occupied by George); and (2) putting Don in a Wednesday cell (since George is already in it). This leaves the following five possible scenarios:

- **Scenario 1** Putting Don in the Tuesday cell, thus assigning Ed to the Thursday cell.
- **Scenario 2** Putting Don in the Thursday cell, thus assigning Ed to the Saturday cell.
- **Scenario 3** Putting Don in the Friday cell, thus assigning Ed to the Sunday cell.
- **Scenario 4** Putting Don in the Saturday cell, thus assigning Ed to the Monday cell.
- **Scenario 5** Putting Don in the Sunday cell, thus assigning Ed to the Tuesday cell.

These scenarios can be shown in the chart as follows:

	MONDAY	TUESDAY	WEDNESDAY	THURSDAY	FRIDAY	SATURDAY	SUNDAY
SCENARIO 1		DON	GEORGE	ED			
SCENARIO 2			GEORGE	DON		ED	
SCENARIO 3			GEORGE		DON		ED
SCENARIO 4	ED		GEORGE			DON	
SCENARIO 5		ED	GEORGE				DON

Clue 1 tells you that *Al's birthday is as many days before George's as Bill's is after Ed's.* In terms of the chart, this means that Al is to be assigned to a cell that is the same number of cells *before* George's cell that Bill's cell is *after* Ed's cell.

Consider scenario 1. As you can see, Al can be put only in the Monday cell, since it is the only empty cell before George. Since Al's birthday falls two days *before* George's, then Bill must be assigned to a cell that is two days *after* Ed's cell—namely, to the Saturday cell:

	MONDAY	TUESDAY	WEDNESDAY	THURSDAY	FRIDAY	SATURDAY	SUNDAY
SCENARIO 1	AL	DON	GEORGE	ED		BILL	
SCENARIO 2			GEORGE	DON		ED	
SCENARIO 3			GEORGE		DON		ED
SCENARIO 4	ED		GEORGE			DON	
SCENARIO 5		ED	GEORGE				DON

But, as the chart shows you, this sequence of birthdays is not consistent with the basic condition set forth by the puzzle—namely, that the birthdays fall on consecutive days. As you can see, there is a gap of one day between Ed's and Bill's birthdays. So, you can discard scenario 1 and move on to scenario 2. In this case, Al can be put into the Monday or Tuesday cell—both of which are empty. If you put Al in the Monday cell, it means that Al's birthday falls two days *before* George's. So, you must put Bill in a cell that is two days *after* Ed's—that is, also in the Monday cell:

	MONDAY	TUESDAY	WEDNESDAY	THURSDAY	FRIDAY	SATURDAY	SUNDAY
SCENARIO 2	AL/BILL		GEORGE	DON		ED	
SCENARIO 3			GEORGE		DON		ED
SCENARIO 4	ED		GEORGE			DON	
SCENARIO 5		ED	GEORGE				DON

But this arrangement also violates the "consecutiveness condition"—that is, that the birthdays fall on consecutive days. So, it too can be discarded. Try putting Al in the Tuesday cell, one day *before* George's birthday. This means that you must put Bill in a cell that is one day *after* Ed's—that is, in the Sunday cell:

	MONDAY	TUESDAY	WEDNESDAY	THURSDAY	FRIDAY	SATURDAY	SUNDAY
SCENARIO 2		AL	GEORGE	DON		ED	BILL
SCENARIO 3			GEORGE		DON		ED
SCENARIO 4	ED		GEORGE			DON	
SCENARIO 5		ED	GEORGE				DON

But the chart reveals that this arrangement is also inconsistent with the "consecutiveness condition" of the puzzle. As you can see, there is a gap of one day between Don's birthday and Ed's birthday. So, scenario 2 can now be safely rejected.

Consider scenario 3. Once again, you can put Al in either the Monday or Tuesday cell. If you put Al in the Monday cell, it means that Al's birthday falls two days *before* George's. So, you must put Bill in a cell that is two days *after* Ed's—that is, in the Tuesday cell:

	Monday	Tuesday	Wednesday	Thursday	Friday	Saturday	Sunday
Scenario 3	Al	Bill	George		Don		Ed
Scenario 4	Ed		George			Don	
Scenario 5		Ed	George				Don

But, as you can see, this scenario too is inconsistent with the "consecutiveness condition" of the puzzle. If you put Al in the Tuesday cell, it means that Al's birthday falls one day *before* George's. So, you must put Bill in a cell that is one day *after* Ed's—that is, in the Monday cell:

	Monday	Tuesday	Wednesday	Thursday	Friday	Saturday	Sunday
Scenario 3	Bill	Al	George		Don		Ed
Scenario 4	Ed		George			Don	
Scenario 5		Ed	George				Don

But, this scenario too is inconsistent with the "consecutiveness condition" of the puzzle. So, you can safely discard scenario 3.

Go on to scenario 4. As you can clearly see, the only cell to which Al can be assigned is the Tuesday cell. This means that Al's birthday falls one day *before* George's. So, you must put Bill in a cell that is one day *after* Ed's—that is, also in the Tuesday cell:

	Monday	Tuesday	Wednesday	Thursday	Friday	Saturday	Sunday
Scenario 4	Ed	Al/Bill	George			Don	
Scenario 5		Ed	George				Don

This scenario is also inconsistent with the "consecutiveness condition" of the puzzle. Once again, you can reject scenario 4 and move on to scenario 5. As you can see, the only cell to which Al can be assigned in this scenario is the Monday cell. This means that Al's birthday falls two days *before* George's. So, you must put Bill in a cell that is two days *after* Ed's—that is, in the Thursday cell:

	Monday	Tuesday	Wednesday	Thursday	Friday	Saturday	Sunday
Scenario 5	Al	Ed	George	Bill			Don

Scenario 5 meets the "consecutiveness condition" of the puzzle. So, the birthdays, in consecutive order, are:

Sunday	→	Don's birthday
Monday	→	Al's birthday
Tuesday	→	Ed's birthday
Wednesday	→	George's birthday
Thursday	→	Bill's birthday

Example 2 The following is a typical example of a clock puzzle.

> Sandra and George made arrangements to meet at the airport to catch the eight o'clock flight to Philadelphia. Sandra thinks that her watch is 25 minutes fast, although it is actually 10 minutes slow. George thinks his watch is 10 minutes slow, while it is actually 5 minutes fast. If both persons rely on their watches, what will happen if they attempt to arrive at the airport 5 minutes before the flight departs?

Consider each person's case separately. When Sandra looks at her watch and sees, say, 8:20, she'll believe that it is 7:55 airport time. Why? Because she thinks that her watch is 25 minutes *fast.* So, in her mind she believes that the time shown by the airport clock is 25 minutes *earlier* than the 8:20 shown by her watch. Thus, *in her mind,* it is 7:55 airport time. Consequently, Sandra believes she has 5 minutes to spare before take-off time at 8:00.

But what time is it, really, when she looks at her watch showing 8:20? In reality, her watch is 10 minutes *slow.* So, the time shown by the airport clock is actually 10 minutes *later* than the 8:20 shown by her watch. Therefore, when her watch shows 8:20, the airport time is actually 8:30. Let's summarize Sandra's case in chart form:

WHAT SANDRA BELIEVES

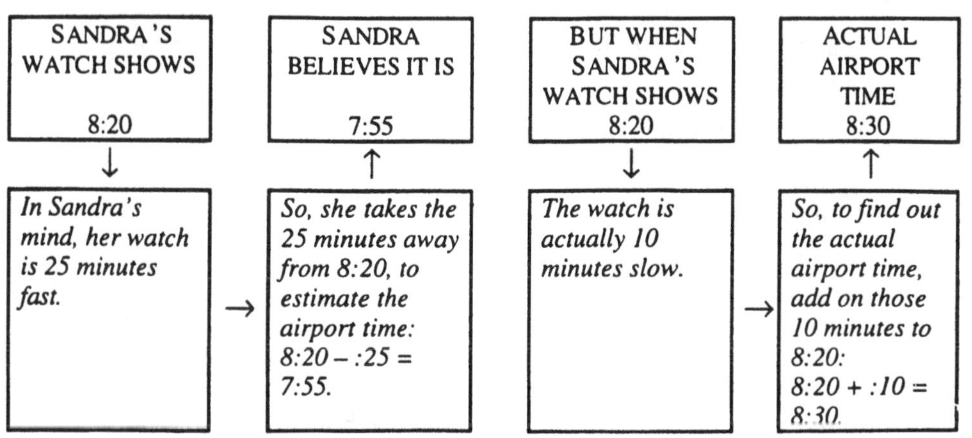

Since the plane is scheduled to leave at 8:00, Sandra will have missed her plane by half an hour!

When George looks at his watch and sees, say, 7:45, what time does he think it is? Since he believes that his watch is 10 minutes *slow*, then he thinks that the time shown by the airport clock is 10 minutes *later* than the 7:45 he sees on his watch. He thinks, therefore, that it is really 7:55, just like Sandra did. So, he too thinks he has 5 minutes to spare before take-off time at 8:00.

But what time is it, really, when he looks at his watch showing 7:45? In reality, his watch is 5 minutes *fast*. So, the time shown by the airport clock is actually 5 minutes *earlier* than the 7:45 time shown by his watch. Therefore, when George's watch shows 7:45, the airport time is actually 7:40.

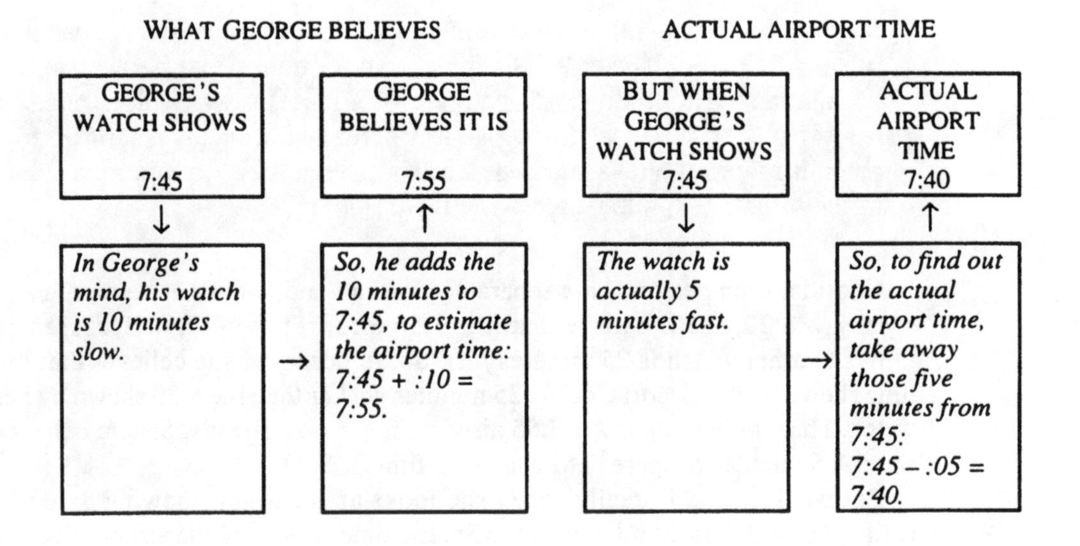

WHAT GEORGE BELIEVES
ACTUAL AIRPORT TIME
GEORGE'S WATCH SHOWS 7:45
GEORGE BELIEVES IT IS 7:55
BUT WHEN GEORGE'S WATCH SHOWS 7:45
ACTUAL AIRPORT TIME 7:40
In George's mind, his watch is 10 minutes slow.
So, he adds the 10 minutes to 7:45, to estimate the airport time: 7:45 + :10 = 7:55.
The watch is actually 5 minutes fast.
So, to find out the actual airport time, take away those five minutes from 7:45: 7:45 – :05 = 7:40.

So, George will actually arrive at the airport at 7:40, 20 minutes ahead of departure time at 8:00.

Summary

The key to solving the type of time puzzle illustrated in example 1 is to draw an appropriate chart showing:

- the relevant time units (days, months, years);
- the sequential order of the units (Monday to Sunday, January to December, etc.);
- a layout of the various hypothetical scenarios that can be inferred from the given information.

In the case of clock puzzles (example 2), it is essential that you employ concrete, step-by-step reasoning, using charts whenever possible. Some of the clock puzzles below will also involve a little algebra.

□□■ Puzzles 67–73

Answers, along with step-by-step solutions, can be found at the end of the chapter.

67 If you add one-quarter of the time from midnight until the present time to half the time from now until midnight, you will get the present time. What time is it?

68 Two weeks ago Daniela bought a grandfather clock that strikes the number of hours every hour. If it takes it 7 seconds to strike 8 o'clock, for how many seconds each day is the clock striking?

69 Becky, Alice, Ted, and Bob all work in the same factory, but their rates of work are different. Becky can do a job in 6 hours. Alice is faster at it, and can do the same job in 5 hours. Ted is faster still, being able to do the job in 4 hours. And Bob is the fastest of the four, since he can do the job in just 3 hours.

Yesterday, Becky started a job alone. After one hour, she was joined by Alice and Ted. An hour after that, Bob joined the other three, and the four completed the job together. How many hours did it take to complete the job, from the time Becky started to work on it?

70 Five couples got married last month, each on a different day of the week, from Monday to Friday. The women's names are Ava, Cynthia, Eartha, Fanny, and Isabella. The men's names are Peter, Rick, Steve, Vic, and Willy.

1 Ava was married on Monday, but not to Willy.

2. Steve got married on Wednesday and Rick on Friday (but not to Isabella).

3. Vic and Fanny got married the day after Eartha's wedding.

Can you determine each couple and the day on which each couple got married?

71 How many minutes after 8 o'clock will the minute hand overtake the hour hand?

72 Georgette works every second day at Hillary's Milk Store as a part-time sales clerk. Alexander also works there as a part-time sales clerk, but every third day. The store stays open seven days a week. This week, Georgette started work on Tuesday, June 1, and Alexander on Wednesday, June 2. On what date will the two be working together?

73 It takes Beatrice twice as long as it takes Amber to do a certain piece of work. Working together, they can do the work in 6 days. How long would it take Amber to do it alone?

Answers and Solutions

 **Answer:** *9:36 A.M.*

Solution: Let x stand for the present time. What does the statement *one-quarter of the time from midnight until the present time* mean in actual hours? A few concrete examples might help you see what it means. If the present time is 9:00 A.M., then, clearly, the time from midnight to 9:00 A.M. is 9 hours. (If you don't see this, just look at your watch!) If the present time is 10:00 A.M., then the time from midnight to 10:00 A.M. is 10 hours. If it is 2:00 P.M., which is 14:00 P.M. according to official time, then the time from midnight to 14:00 P.M. is 14 hours. And so on. So, if the present time is x, then *the time from midnight to* x *is* x hours. And, of course, *one-quarter* of that time is $\frac{1}{4}x$.

What does *half the time from now until midnight* mean in actual hours? Once again, a few concrete examples might help you see what it means. If the present time is 9:00 A.M., then *the time from now until midnight* is 15 hours. Using 24:00 as the midnight hour, you get this figure simply by subtracting 24:00 – 9:00 = 15 hours. Similarly, if it is 14:00 P.M., then *the time from now until midnight* is 24:00 – 14:00 = 10 hours. And so on. So, if the present time is x, then *the time from x until midnight* is $24 - x$. And, of course, *half* that time is $\frac{1}{2}(24 - x)$.

Now, the puzzle says that if you *add* $\frac{1}{4}x$ to $\frac{1}{2}(24 - x)$, you will get the present time:

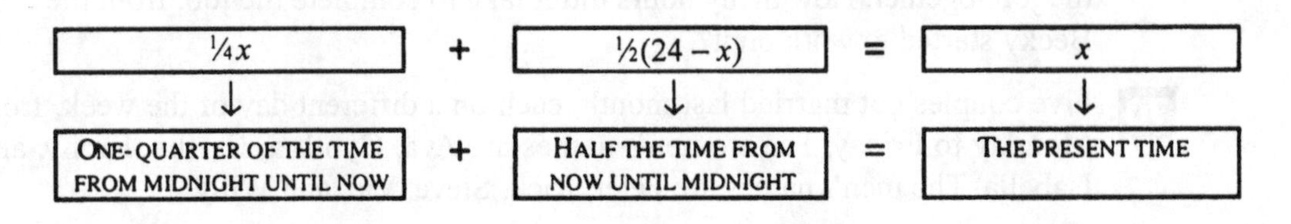

Solving for x, you'll get: $x = 9.6$ hours. What's .6 of an hour? There are 60 minutes in an hour, so .6 of 60 minutes is 36 minutes. Therefore, the time is 9:36 A.M.

68 **Answer:** *132 seconds.*

Solution: You are told that the clock takes 7 seconds to strike 8 o'clock. Let's see what that actually means. The clock will give out eight strikes when it strikes 8 o'clock. These eight strikes are separated by seven equal intervals of time. A simple diagram will help you visualize this in concrete terms:

STRIKE 1 ↓ | STRIKE 2 ↓ | STRIKE 3 ↓ | STRIKE 4 ↓ | STRIKE 5 ↓ | STRIKE 6 ↓ | STRIKE 7 ↓ | STRIKE 8 ↓

INTERVAL 1 | INTERVAL 2 | INTERVAL 3 | INTERVAL 4 | INTERVAL 5 | INTERVAL 6 | INTERVAL 7

The puzzle tells you that to make the eight strikes the clock takes 7 seconds. This means, therefore, that each of the seven intervals is 1 second long. This applies, of course, to every hour on the hour. For example, at 9 o'clock, the clock will make nine strikes, separated by eight intervals lasting 1 second each. So, it will take the clock 8 seconds to strike 9 o'clock. At 10 o'clock the clock will make ten strikes separated by nine intervals lasting 1 second each. So, it will take the clock 9 seconds to strike 10 o'clock. And so on. In general, at each hour the clock will take 1 second less than the number of the hour to strike that hour.

Make an appropriate chart to show the relation between the number of strikes and the time taken to strike each hour from 1:00 to 12:00:

Hour	Strikes	Time Taken
1:00	1	0 sec.
2:00	2	1 sec.
3:00	3	2 sec.
...	...	...
12:00	12	11 sec.

To find out how many seconds the clock has been striking from 1:00 to 12:00, all you have to do is to add up the numbers in the *Time Taken* column. These are, as you can see, the first 11 integers: $0+1+2+3+4+5+6+7+8+9+10+11=66$ seconds. Now, since the clock goes around two times a day from 1:00 A.M. to 12:00 noon and from 1:00 P.M. to 12:00 midnight, the number of seconds the clock will be striking is twice this number: $66 \times 2 = 132$ seconds per day.

69 **Answer:** *$2\frac{13}{57}$ hours.*

Solution: This puzzle requires that a common standard against which to compare the working rates of the four individuals be set up. The most convenient standard is, of course, the *per hour* one.

Start with Becky. You are told that she completes a job in 6 hours. So, every hour she completes $\frac{1}{6}$ of that job. This can be easily confirmed as follows. She completes $\frac{1}{6}$ of the job in the first hour, $\frac{2}{6}$ ($= \frac{1}{6} + \frac{1}{6}$) in the second hour, and so on. During the sixth hour, she completes $\frac{6}{6}$ of the job, which, of course, is equal to 1 (the whole job).

With similar reasoning, you can calculate the rates of the others easily. Alice completes a job in 5 hours, so she completes $\frac{1}{5}$ of the job every hour. Ted completes it in 4 hours, so he completes $\frac{1}{4}$ of the job every hour. Bob completes the job in 3 hours, so he completes $\frac{1}{3}$ of the job per hour:

PROPORTION OF THE JOB COMPLETED PER HOUR BY . . .

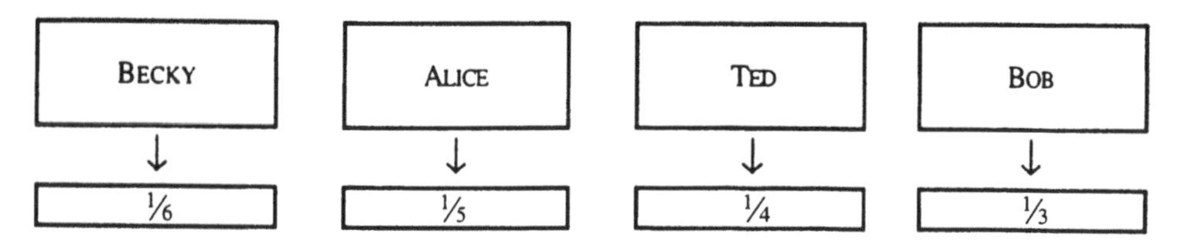

Now, you are asked to find out how many hours a certain job took to complete. Represent the number of hours it took with x. Becky worked the whole x hours. Alice and Ted joined Becky an hour after Becky started. So, they worked 1 hour less than her x hours. Algebraically, this translates, of course, into

$x - 1$ hours. Finally, Bob joined the group 2 hours after Becky started the job—because 1 hour after Alice and Ted is 2 hours after Becky. So, Bob worked for $x - 2$ hours:

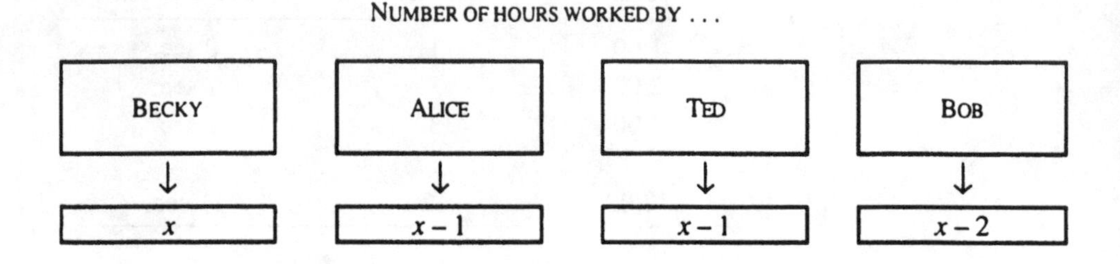

Now, let's see how much of the job each one completed in one hour. Becky's completion rate is $\frac{1}{6}$ of the job per hour. So, in x hours she completed $\frac{1}{6}x$ of the job. Alice's rate is $\frac{1}{5}$ of the job per hour. She worked $x - 1$ hours. So, she completed $\frac{1}{5}(x - 1)$ of the job. Ted's completion rate is $\frac{1}{4}$ of the job per hour. He worked $x - 1$ hours, so his share of the load came to $\frac{1}{4}(x - 1)$ of the job. Finally, Bob's completion rate is $\frac{1}{3}$ of the job per hour. He worked $x - 2$ hours on this job; so he completed $\frac{1}{3}(x - 2)$ of the job.

To summarize:

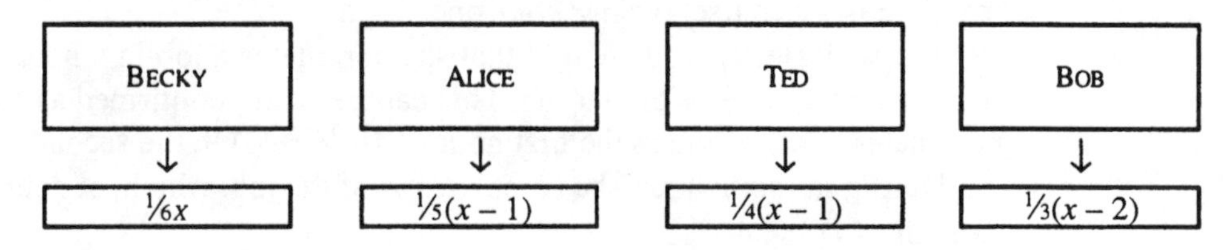

The four individuals, working at their normal paces, completed the job together. This is equivalent to saying that the sum of the fractions of the job each one completed equals the whole job, or 1:

$$\frac{1}{6}x + \frac{1}{5}(x - 1) + \frac{1}{4}(x - 1) + \frac{1}{3}(x - 2) = 1$$

Solving for x, you'll get: $x = 2\frac{13}{57}$. So, it took $2\frac{13}{57}$ hours for the four of them to complete the job working together, as stipulated.

70 **Answer:** *Peter and Ava got married on Monday, Willy and Isabella on Tuesday, Steve and Eartha on Wednesday, Vic and Fanny on Thursday, and Rick and Cynthia on Friday.*

Solution: This puzzle combines elements of deductive and time logic. First, set up a two-axis chart for correlating the names of the two individuals in each couple:

	Ava	Cynthia	Eartha	Fanny	Isabella
Peter					
Rick					
Steve					
Vic					
Willy					

Then, set up a time chart for registering the days on which the marriages took place:

Couple	Monday	Tuesday	Wednesday	Thursday	Friday
Husband					
Wife					

Start by inserting Ava into the Monday cell in the time chart, because statement 1 tells you that she was married on Monday:

Couple	Monday	Tuesday	Wednesday	Thursday	Friday
Husband					
Wife	Ava				

Statement 1 also tells you that Ava did not marry Willy. Show this in the couples chart in the usual way:

	Ava	Cynthia	Eartha	Fanny	Isabella
Peter					
Rick					
Steve					
Vic					
Willy	×				

Statement 2 tells you that Steve got married on Wednesday and Rick on Friday. Register these facts in the time chart:

Couple	Monday	Tuesday	Wednesday	Thursday	Friday
Husband			Steve		Rick
Wife	Ava				

From this chart, you can see that neither Steve nor Rick married Ava (whose marriage was on Monday). Statement 2 also tells you that Rick did not marry Isabella. Show all these new facts in the couples chart:

	Ava	Cynthia	Eartha	Fanny	Isabella
Peter					
Rick	×				×
Steve	×				
Vic					
Willy	×				

Statement 3 tells you that Vic and Fanny make up a couple:

	Ava	Cynthia	Eartha	Fanny	Isabella
Peter				×	
Rick	×			×	×
Steve	×			×	
Vic	×	×	×	●	×
Willy	×			×	

The couples chart now reveals that Ava and Peter make up another couple, since the only cell left under Ava is opposite Peter:

	Ava	Cynthia	Eartha	Fanny	Isabella
Peter	●	×	×	×	×
Rick	×			×	×
Steve	×			×	
Vic	×	×	×	●	×
Willy	×			×	

Register this new finding as well in the time chart, putting Peter in the Monday cell with Ava:

Couple	Monday	Tuesday	Wednesday	Thursday	Friday
Husband	Peter		Steve		Rick
Wife	Ava				

Vic and Fanny could not have married on Tuesday because they married after Eartha (statement 3), and since Eartha did not marry on Monday (because Ava did), the earliest she could have married was on Tuesday. Vic and Fanny did not marry on Wednesday because Steve and his spouse did; nor did they marry on Friday, because Rick and his spouse did. So they got married on Thursday, the only day left for them in the time chart:

Couple	Monday	Tuesday	Wednesday	Thursday	Friday
Husband	Peter		Steve	Vic	Rick
Wife	Ava			Fanny	

The above chart now reveals that the only day left for Willy is Tuesday:

Couple	Monday	Tuesday	Wednesday	Thursday	Friday
Husband	Peter	Willy	Steve	Vic	Rick
Wife	Ava			Fanny	

Eartha could not have gotten married on Friday to Rick, because she got married before Vic and Fanny did. So Rick and Eartha are not a couple:

	Ava	Cynthia	Eartha	Fanny	Isabella
Peter	●	×	×	×	×
Rick	×		×	×	×
Steve	×			×	
Vic	×	×	×	●	×
Willy	×			×	

The couples chart now reveals that Rick married Cynthia, because the only cell left opposite Rick is under Cynthia.:

	Ava	Cynthia	Eartha	Fanny	Isabella
Peter	●	×	×	×	×
Rick	×	●	×	×	×
Steve	×	×		×	
Vic	×	×	×	●	×
Willy	×	×		×	

This means, of course, that Cynthia, being Rick's spouse, got married on Friday:

Couple	Monday	Tuesday	Wednesday	Thursday	Friday
Husband	Peter	Willy	Steve	Vic	Rick
Wife	Ava			Fanny	Cynthia

Now, according to statement 3, Vic and Fanny got married the day after Eartha did. So, Eartha got married on Wednesday:

Couple	Monday	Tuesday	Wednesday	Thursday	Friday
Husband	Peter	Willy	Steve	Vic	Rick
Wife	Ava		Eartha	Fanny	Cynthia

As you can see, this means that she married Steve:

	AVA	CYNTHIA	EARTHA	FANNY	ISABELLA
PETER	●	×	×	×	×
RICK	×	●	×	×	×
STEVE	×	×	●	×	×
VIC	×	×	×	●	×
WILLY	×	×	×	×	

This leaves Willy and Isabella as the couple who got married on Tuesday:

	AVA	CYNTHIA	EARTHA	FANNY	ISABELLA
PETER	●	×	×	×	×
RICK	×	●	×	×	×
STEVE	×	×	●	×	×
VIC	×	×	×	●	×
WILLY	×	×	×	×	●

COUPLE	MONDAY	TUESDAY	WEDNESDAY	THURSDAY	FRIDAY
HUSBAND	PETER	WILLY	STEVE	VIC	RICK
WIFE	AVA	ISABELLA	EARTHA	FANNY	CYNTHIA

To summarize, Peter and Ava got married on Monday, Willy and Isabella on Tuesday, Steve and Eartha on Wednesday, Vic and Fanny on Thursday, and Rick and Cynthia on Friday.

71 **Answer:** *$43\frac{7}{11}$ minutes.*

Solution: Let x represent the number of minutes (=divisions on a clock) after 8 o'clock required by the minute hand to overtake the hour hand. When it is precisely 8 o'clock, the hour hand is at the 8 o'clock division point and the minute hand at the 12 o'clock one. The hour hand will be at the x division point after 8:00 when the minute hand overtakes it. At precisely 8 o'clock there are 40 divisions (=40 minutes) separating the minute hand from the hour hand. So, to overtake the hour hand, the minute hand will have to cover the 40 divisions to the 8 o'clock division point plus the extra x divisions after 8:00 that the hour hand has covered. Altogether, it will have to cover a distance of $40 + x$ divisions (=minutes) when it overtakes the hour hand. The hour hand, of course, will have covered a distance equal to just those x divisions (=minutes). If you have difficulty following this line of reasoning, go over this paragraph a few times, drawing a clock diagram and marking it up as described in the paragraph.

Remember that $D = R \times T$ and, therefore, that $R = D/T$. So, go ahead and show the travel rates to division point x after 8:00 for both hands:

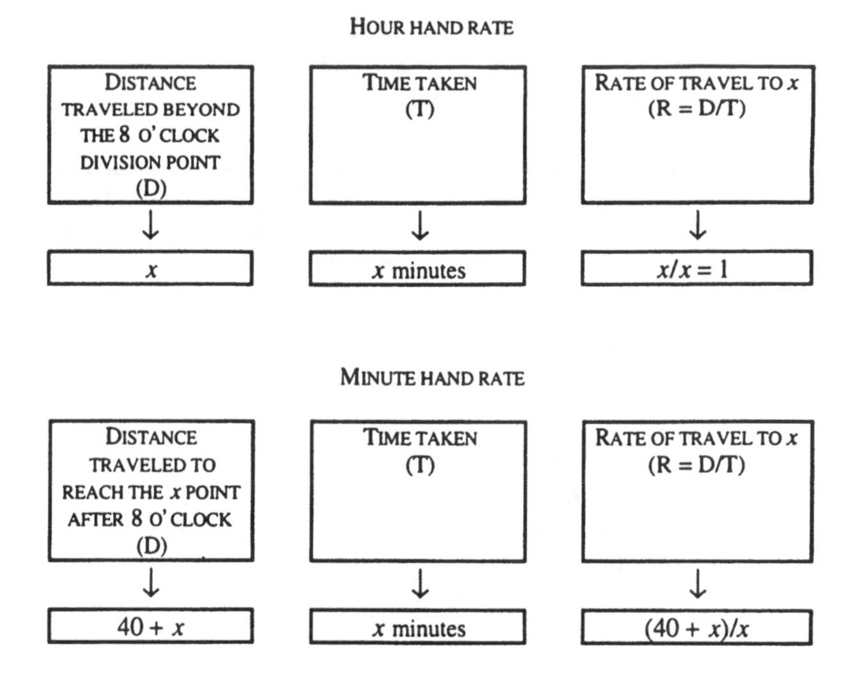

A minute hand must cover 60 minutes, or 60 divisions on the dial, from one hour to the next hour. An hour hand, on the other hand, must cover 5 such divisions in the same time. Consequently, an hour hand moves at $\frac{5}{60}$, or $\frac{1}{12}$, of the rate moved by the minute hand in one hour. Therefore, since the hour hand moves at $\frac{1}{12}$ the rate of the minute hand, it traveled the distance to x in the same time, but at $\frac{1}{12}$ the rate of the minute hand.

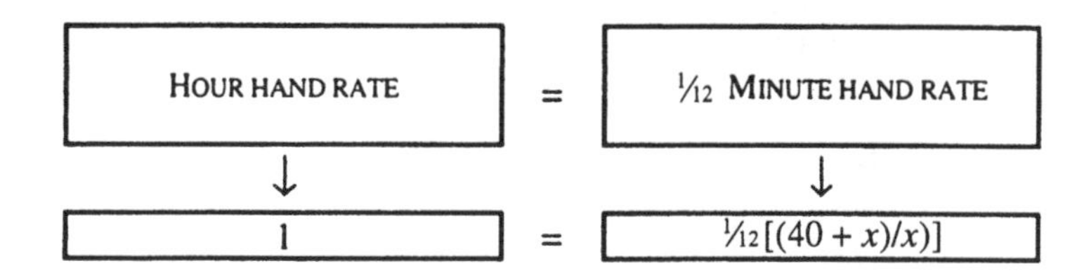

Solving for x, you'll get: $x = 3\frac{7}{11}$. So, $3\frac{7}{11}$ minutes after 8:00, the minute hand will overtake the hour hand. This means that it will take the 40 minutes from the 12 o'clock point to the 8 o'clock point plus the $3\frac{7}{11}$ divisions moved by the hour hand for the minute hand to overtake the hour hand. In total, therefore, it will take $43\frac{7}{11}$ minutes.

72 **Answer:** *Saturday, June 5.*

Solution: Since Tuesday is June 1, then Monday is May 31. This suggests a time chart such as the following one:

CLERK	MONDAY, MAY 31	TUESDAY, JUNE 1	WEDNESDAY, JUNE 2	THURSDAY, JUNE 3	FRIDAY, JUNE 4	SATURDAY, JUNE 5	SUNDAY, JUNE 6
GEORGETTE							
ALEXANDER							

Since Georgette works every second day and Alexander every third, you can show the layout of Alexander's (A) and Georgette's (G) working schedules for the week of May 31 as follows:

Clerk	Monday, May 31	Tuesday, June 1	Wednesday, June 2	Thursday, June 3	Friday, June 4	Saturday, June 5	Sunday, June 6
Georgette		G		G		G	
Alexander			A			A	

The chart reveals that the two will be working together on Saturday, June 5.

73 **Answer:** *9 days.*

Solution: This is really a kind of opposite to puzzle 69 above. Start by letting x represent the number of days it takes Amber to do the work alone. Since it takes Beatrice twice as long as Amber to do a job, she would thus take $2x$ days to finish it.

Now, follow the same line of reasoning that was employed to solve puzzle 69. Since it takes Amber x days to complete the job alone, she can be seen to complete $1/x$ of the job in one day. And since it takes Beatrice $2x$ days to complete the job, she can be seen to complete $\frac{1}{2}x$ of the job in one day. Working together, they can complete $1/x + \frac{1}{2}x$ of the job in one day.

The job takes 6 days working together. So, $\frac{1}{6}$ of the job gets completed by the two working together every day:

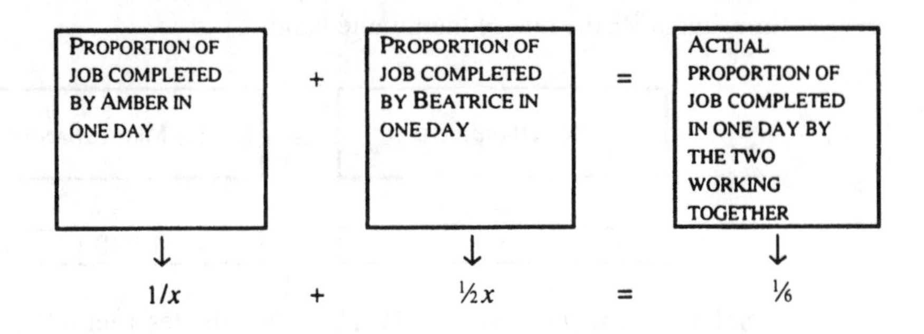

Solving for x, you'll get: $x = 9$. So, it would take Amber 9 days to complete the job alone.

10 Puzzles in Paradox Logic

In logic and mathematics, a *paradox* is defined as an apparently contradictory conclusion that is derived from what seem to be valid premises. Paradoxes have been known since the time of the Greek philosopher Zeno of Elea in the fifth century B.C. On inspection, many paradoxes prove to be based on false premises or arguments, on incomplete presuppositions, or on the structure of language.

The following paradox is known by virtually everyone: *Which came first, the chicken or the egg?* Like many paradoxes, it entails a logical *circularity,* that is, reasoning that goes around in circles. If I say that "the chicken came first," then you would counter with "Impossible, the chicken had to hatch from an egg!" "So," I would reply, "it would seem that the egg came first." "Impossible again," you would retort. "The egg had to be laid first by a chicken." And in this way our conversation would continue, going around and around in circles forever!

How To...

Puzzles in paradox logic require, above all else, a close and attentive reading of the puzzle's language. In this chapter you will encounter *authentic paradoxes* of various kinds, but also those that on the surface seem to present you with a paradox but in reality are simple puzzles in logical deduction (*apparent paradoxes*).

Puzzle Properties

There are two main types of *authentic paradoxes:* (1) those that entail a logical circularity, and (2) those that arise because of a logical inconsistency or fallacy. Your task is to flesh out the circularity or the inconsistency and expose it for what it is. These puzzles will teach you, therefore, the value of grasping the meaning of a statement, or of a logical demonstration, in its essence.

Apparent paradoxes are puzzles which at first reading appear to pose a paradoxical enigma but which in actual fact can be solved in a straightforward manner. In other puzzle books, you will find that these puzzles have been classified under a different rubric. The decision to include them in this chapter has been motivated simply by the fact that they *appear* to be paradoxical. That *in itself* is well within the nature of paradox logic!

As you might expect, there are no pre-established techniques for solving such puzzles. A few examples will suffice to show you what is involved. These can be used as "templates" for solving similar puzzles in the future.

Example 1 A classic example of logical circularity is the liar paradox, formulated by the ancient Greek poet Epimenides in the sixth century B.C. It has come down to us more or less in the following form:

> The Cretan philosopher Epimenides once said: "All Cretans are liars." Did Epimenides speak the truth?

Because Epimenides is himself a Cretan, his statement leads to a circularity. Here's why. Epimenides is a Cretan. Therefore, he is a liar. So, his statement is false. What is his statement? *All Cretans are liars.* But, wait! The statement is true! Cretans are indeed all liars. So, Epimenides is not a liar after all. He told it like it is. But, then, his statement leads us to conclude that Epimenides is both a liar and not a liar, which is logically impossible!

Another kind of paradox puzzle is one that seems to show the existence of a contradiction or an impossibility but in reality is based on either a false assumption or an unpermitted procedure in logic. Here's an example of such a puzzle:

Assume that:	(1) $a = b$.
Multiply both sides by a:	(2) $a^2 = ab$.
Subtract b^2 from both sides:	(3) $a^2 - b^2 = ab - b^2$.
Factor both sides:	(4) $(a + b)(a - b) = b(a - b)$.
Divide both sides by $(a - b)$:	(5) $a + b = b$.
We started off by assuming $a = b$, so:	(6) $b + b = b$.
Therefore:	(7) $2b = b$.
Or:	(8) $2 = 1$.

We have proven that 2 = 1. Or have we?

The contradiction in this demonstration arises because it uses an unpermitted procedure. It started off by assuming $a = b$. This means that $a - b = 0$. When the equation $(a + b)(a - b) = b(a - b)$ was divided by $a - b$ in step (5) above, the equation was, consequently, divided by 0 and this, as you might recall from your school algebra, is not a permitted operation. Indeed, you can now understand why this is so!

Example 2 *Apparent paradoxes* are loosely classified as puzzles in paradox logic because something in their statement, or in their solution, seems paradoxical but in reality is not. One common type concerns kinship relations. Here's an example in this genre.

> A young boy answers the phone. "Who is speaking?" he asks. The voice of a man answers: "Don't you recognize me? Your mother's mother is my mother-in-law." Is that possible?

The man seems to have posed a paradoxical situation to the boy. But the key to solving this puzzle is to examine each of the man's two statements—(1) *your mother's mother* and (2) *is my mother-in law*—concretely with a flow chart:

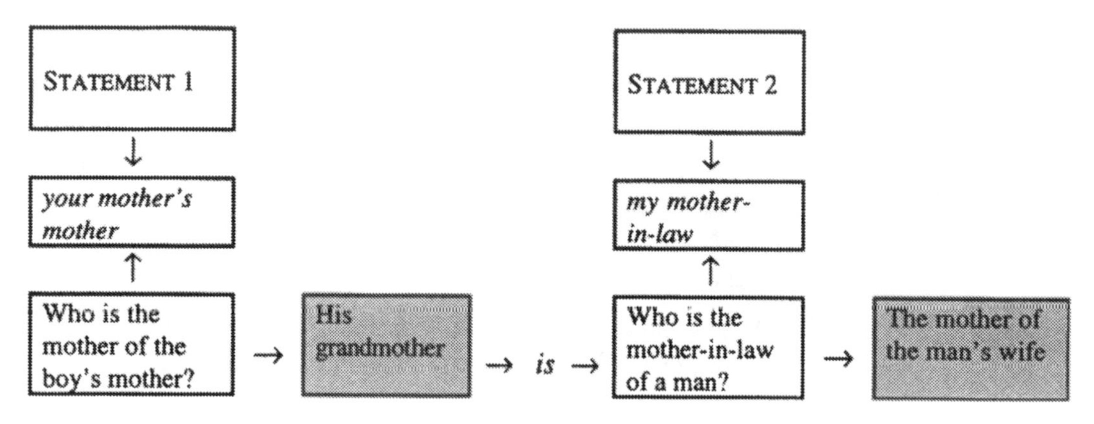

Now, the mother of any man's wife is a grandmother to the man's child. If you do not see this, think of your own grandmother on your mother's side. She is or was, of course, also the mother-in-law of your father, that is, the mother of your father's wife (your mother). So, the man on the phone can only be the boy's father. There is nothing paradoxical about what he says!

Summary

Clearly, the types of puzzle illustrated in examples 1 and 2 cannot be approached systematically in the same way that, say, puzzles in logical deduction, truth logic, algebraic logic, or code logic can. As a general rule, it is useful to write out a puzzle for yourself, thinking about what it says, word by word. This increases your chances of fleshing out a logical inconsistency, an unpermitted logical procedure, and so on.

Puzzles 74–85

Answers, along with step-by-step solutions, can be found at the end of the chapter.

Yesterday I picked up a note handwritten by my spouse. The note said: "*This sentence is false.*" Is the sentence true?

75 In a small village, a barber shaves everyone who does not shave himself. Does he shave himself? Incidentally, this puzzle has come to be known as the *barber paradox*. It was formulated by the twentieth century British philosopher Bertrand Russell (1872–1970).

76 My wily friend showed me a truly enigmatic card last week. On one side it read: "*The sentence on the other side of this card is true.*" On the other it read: "*The sentence on the other side of this card is false.*" Can you tell me which of the statements is true, if any?

77 A loquacious man looks at the portrait of a boy and says, in his typically poetic style: "*Brothers and sisters have I none, but that boy's father is my father's son.*" Who is the boy in the picture?

78 Examine the following algebraic proof.

You are given that:	(1) $a = b + c\ (c \neq 0)$
Multiply both sides by (a – b):	(2) $a(a - b) = b(a - b) + c(a - b)$
The result is:	(3) $a^2 - ab = ab - b^2 + ac - bc$
Simplify the expression:	(4) $a^2 - ab - ac = ab - b^2 - bc$
Factor both sides:	(5) $a(a - b - c) = b(a - b - c)$
Divide both sides by (a – b – c):	(6) $a = b$

Expression (6) now shows that a is equal to b. But in expression (1), a is equal to b plus c. It is thus greater than b. Can you explain this contradiction?

79 Here's a classic from the pen of Lewis Carroll (1832–1898), known for his two great children's novels, *Alice's Adventures in Wonderland* and *Through the Looking Glass*. Which clock keeps the best time? The clock that loses a minute a day or one that doesn't run at all?

A rare gold coin is in one of the following three boxes, each of which has an inscription written on it as follows:

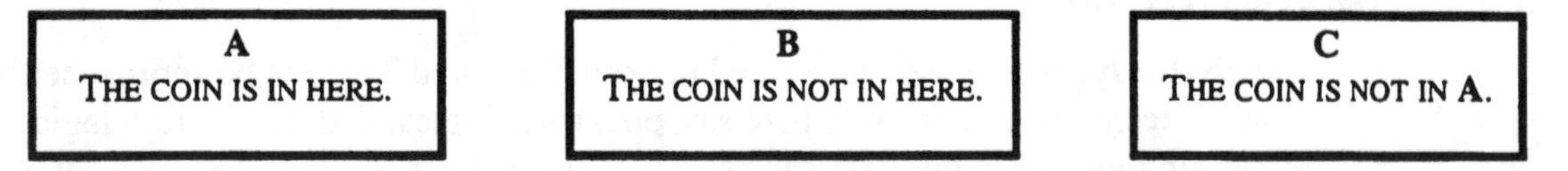

A
THE COIN IS IN HERE.
B
THE COIN IS NOT IN HERE.
C
THE COIN IS NOT IN A.

Can you tell where the coin is if, *at most*, only one of the inscriptions is true?

A year ago at an antique sale I bought a jewelry box that bears the following inscription:

THIS BOX WAS NOT MADE BY
A TRUTH-TELLER.

Was the box made by a truth-teller or a liar?

82 Let's assume that: $x + y = y$. Now, let's assign some values to x and y in this equation.

If $x = 0$ and $y = 1$, we get $0 + 1 = 1$, which is, of course, correct.
If $x = 1$ and $y = 2$, we get $1 + 2 = 2$, which is incorrect.

How is this so?

83 This sentence is not false. Is it true or false?

84 My sister never tells me the truth. So, what am I to make of what she said to me yesterday: "*I have never told you the truth*"?

85 Here's an appropriate, and paradoxical, way to end this course. I'm lying when I say that you have not become a great puzzle solver. Have you become a great puzzle solver?

Answers and Solutions

74 **Answer:** *The sentence leads to a circularity.*

Solution: If the sentence is true, then what it says—"*This sentence is false*"—is true. What's the upshot of this? Well, if it is true that *this sentence is false*, then the sentence is false (as it asserts). But that would mean that the sentence is both true (= premise) and false (= conclusion)!

So, let's try the opposite premise, namely that the sentence is false. What's the upshot of this new premise? Well, if it is false that *this sentence is false*, then the sentence is true (contrary to what it asserts). But then again this would mean that it is both false (= premise) and true (= conclusion)!

75 **Answer:** *It cannot be determined because the assertion made leads to a circularity.*

Solution: If the barber goes ahead and shaves himself, then he has shaved someone in the village who does, in fact, shave himself—namely *himself,* the barber! If he does not shave himself, then he is leaving out someone in the village who does not shave himself—again, *himself,* the barber!

76 **Answer:** *The truth value of the statements cannot be determined, because they lead to a circularity.*

Solution: For the sake of concreteness, the two sides of the card can be labeled **A** and **B** and rephrased as follows:

A: *The sentence on the other side of this card is true*
↓
The sentence on **B** *is true*

B: *The sentence on the other side of this card is false*
↓
The sentence on **A** *is false*

Consider the statement on **A**. The two premises that it entails both lead to contradictory conclusions as follows:

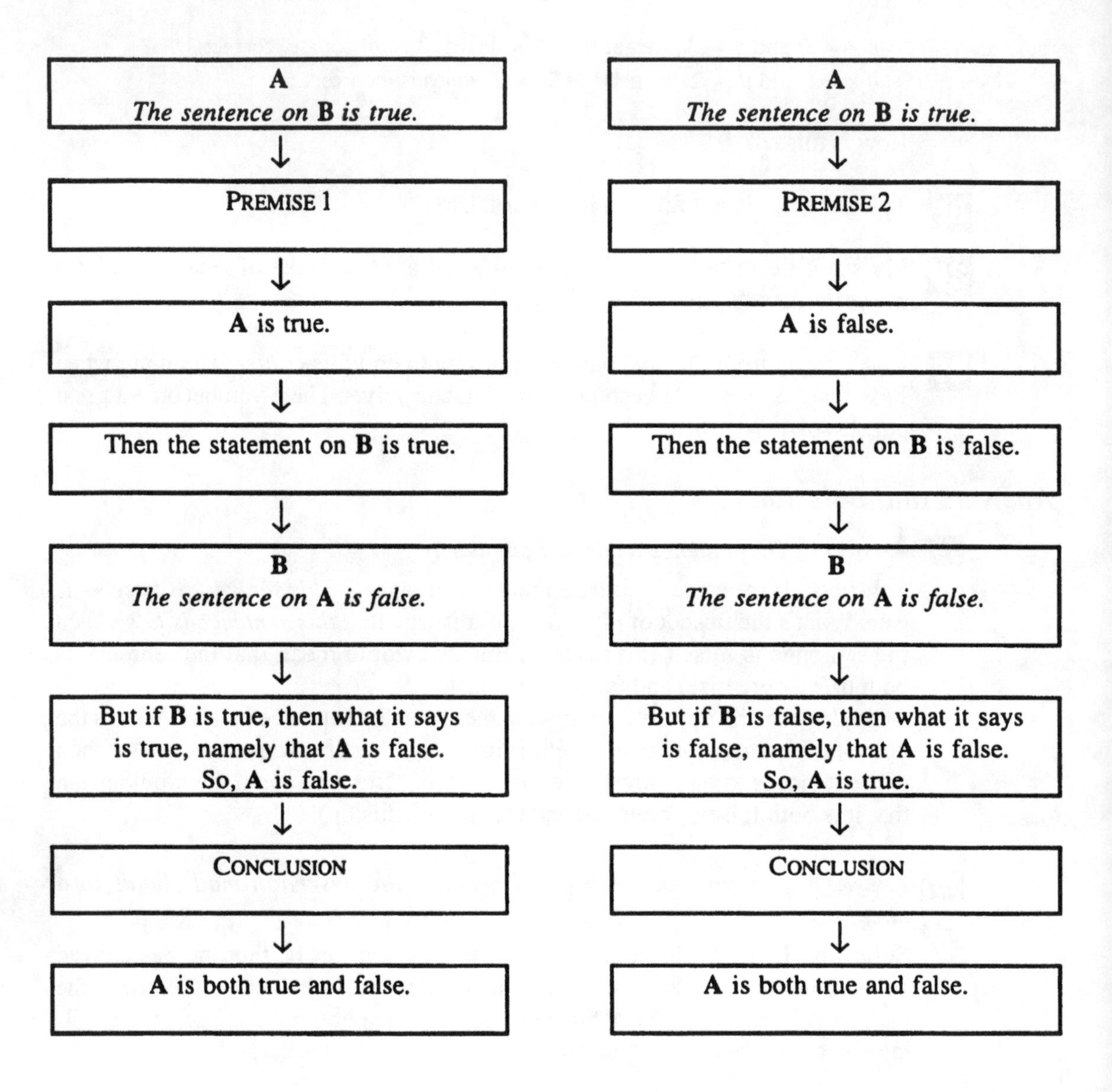

A
The sentence on B is true.
PREMISE 1
A is true.
Then the statement on B is true.
B
The sentence on A is false.
But if B is true, then what it says is true, namely that A is false. So, A is false.
CONCLUSION
A is both true and false.
A
The sentence on B is true.
PREMISE 2
A is false.
Then the statement on B is false.
B
The sentence on A is false.
But if B is false, then what it says is false, namely that A is false. So, A is true.
CONCLUSION
A is both true and false.

The same contradictory conclusions result by considering the statement on **B** in the same way.

77 **Answer:** *The son of the man looking at the picture.*

Solution: Many people get the wrong answer when they first attempt to solve this puzzle, but there is nothing really paradoxical about what it says. It is an example of an *apparent paradox*. For the sake of convenience, let's call the man looking at the picture **A** and the boy in the picture **B**. Let's break down **A**'s statement into separate assertions, drawing appropriate conclusions:

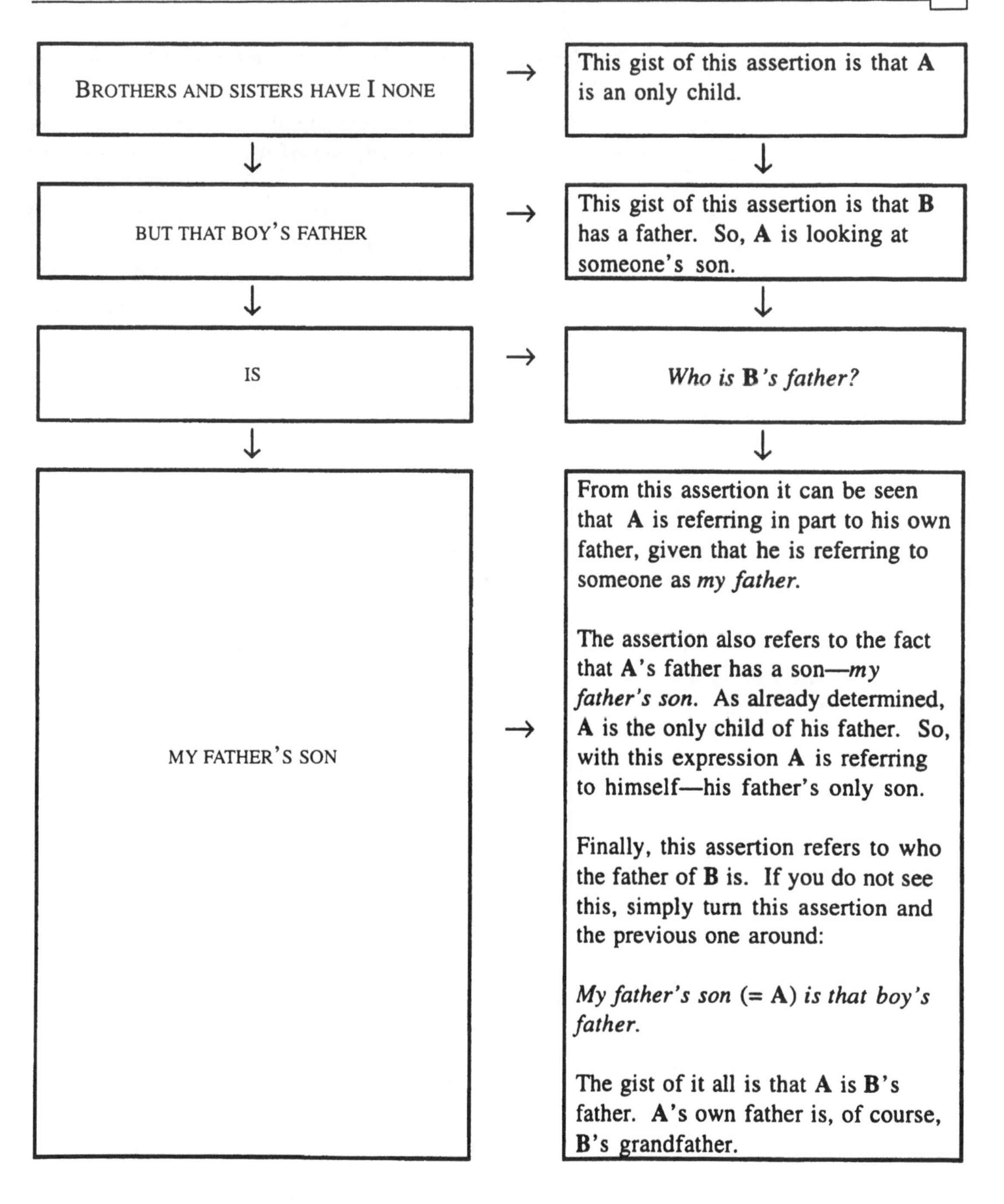

BROTHERS AND SISTERS HAVE I NONE
This gist of this assertion is that **A** is an only child.
BUT THAT BOY'S FATHER
This gist of this assertion is that **B** has a father. So, **A** is looking at someone's son.
IS
Who is **B**'*s father?*
MY FATHER'S SON
From this assertion it can be seen that **A** is referring in part to his own father, given that he is referring to someone as *my father.*
The assertion also refers to the fact that **A**'s father has a son—*my father's son.* As already determined, **A** is the only child of his father. So, with this expression **A** is referring to himself—his father's only son.
Finally, this assertion refers to who the father of **B** is. If you do not see this, simply turn this assertion and the previous one around:
My father's son (= **A**) *is that boy's father.*
The gist of it all is that **A** is **B**'s father. **A**'s own father is, of course, **B**'s grandfather.

78 **Answer:** *In (6)* $a - b - c = 0$ *was used as a divisor.*

Solution: Consider (1): $a = b + c$. This can, of course, be rewritten as $a - b - c = 0$. The flaw in the proof thus occurs in (6) when $a - b - c$, or 0, was used as a divisor. Needless to say, division by 0 is not permitted, leading, as it does, to such contradictory proofs.

79 **Answer:** *According to Carroll, the clock that doesn't run at all keeps the best time.*

Solution: Here's how Lewis Carroll—who was actually Charles L. Dodgson, instructor of mathematics at Christ Church, one of the colleges of Oxford University—reasoned.

Compare the clock that loses one minute per day to a good clock. At 12:00 midnight of the first day, the two clocks are synchronized. After that first day, the bad clock will be off by one minute at midnight—showing 11:59. After the second day, it will be off by two minutes at midnight—showing 11:58. And so on:

Day	Good Clock	Bad Clock
1st	12:00	11:59
2nd	12:00	11:58
3rd	12:00	11:57
...	...	...
60th	12:00	11:00

So, after the 60th day, the bad clock will be off by one hour at midnight—showing 11:00. Therefore, it will be off by another hour after the next 60 days, that is, after 120 days in total:

Day	Good Clock	Bad Clock
61st	12:00	10:59
62nd	12:00	10:58
63rd	12:00	10:57
...	...	...
120th	12:00	10:00

The bad clock will have to lose 12 hours in order to become synchronized once again with the good clock, that is, to show 12:00 correctly. Since each hour it loses takes 60 days, it will need $60 \times 12 = 720$ days to become synchronized again. That is almost two years.

So, Carroll concluded, it would take two years for the clock that loses one minute a day to show the correct time. The stopped clock, on the other hand, shows 12:00 correctly twice a day—at noon and at midnight! On the basis of this kind of reasoning, the stopped clock would appear to be the better one.

80 **Answer:** *The coin is in B.*

Solution: This is another kind of *apparent paradox*. Its statements only resemble contradictory statements such as the ones you have been examining above.

But in actual fact it is a puzzle that can be solved like any puzzle in truth logic (Chapter 2). So, start by assuming that the inscription on **A** is true:

SCENARIO 1

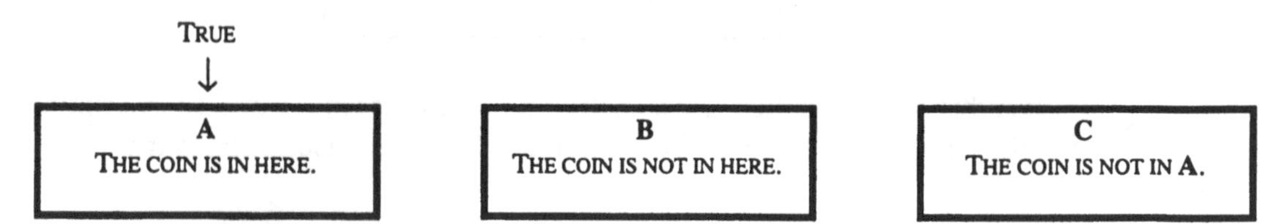

Now, you can quickly ascertain that **B**'s inscription is also true—if the coin is in **A**, then, as **B**'s inscription proclaims, it is certainly not in **B**. But this is contrary to the condition that at most one inscription is true. So, you can reject scenario 1. In the process, however, you have discovered that **A**'s inscription is false—the coin is not in **A**. That makes **C**'s inscription true, since it merely confirms that the coin is not in **A**:

SCENARIO 2

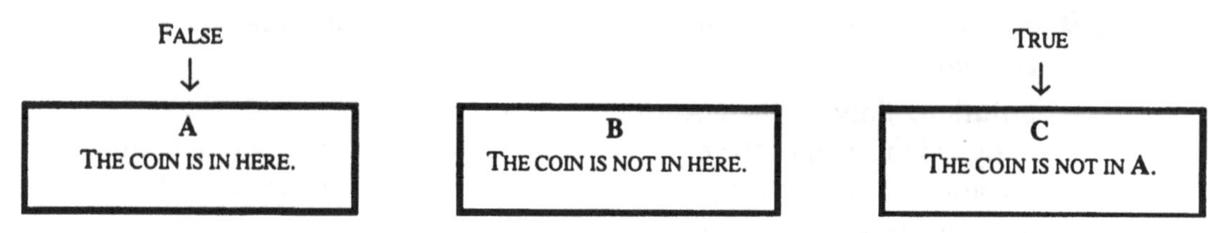

Since at most only one of the inscriptions is true, then **B**'s inscription is false. This completes scenario 2:

SCENARIO 2

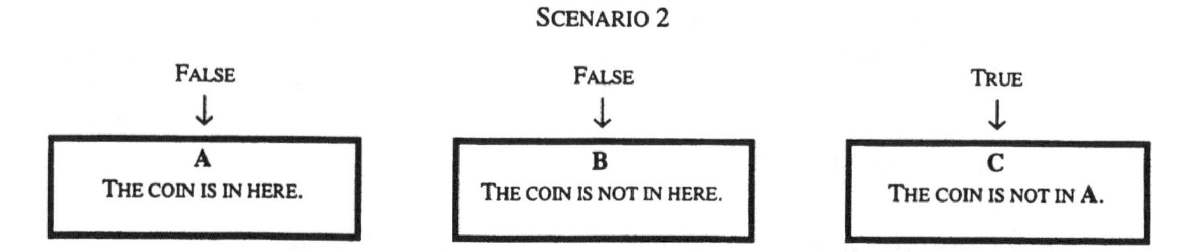

B's inscription reads: "*The coin is not in here.*" According to scenario 2, this is a false statement. Since this F-value does not lead to any contradiction in terms of the entire scenario, it can be safely concluded that the coin is in **B**—contrary to what **B**'s inscription says.

81 **Answer:** *It is not possible to determine who made the box.*

Solution: Assume that the person who made the box was a truth-teller. Then, the inscription is false—since the box says that it was not made by a truth-teller:

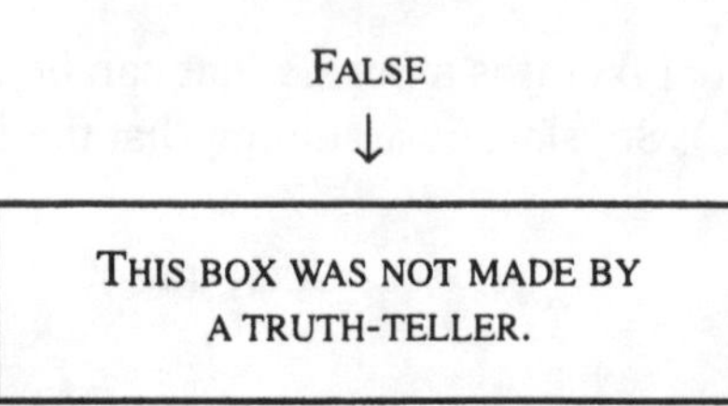
FALSE
↓
THIS BOX WAS NOT MADE BY
A TRUTH-TELLER.

But that cannot be, because a truth-teller would not make a false inscription. So, the person who made the box was a liar. Then, the inscription is true.

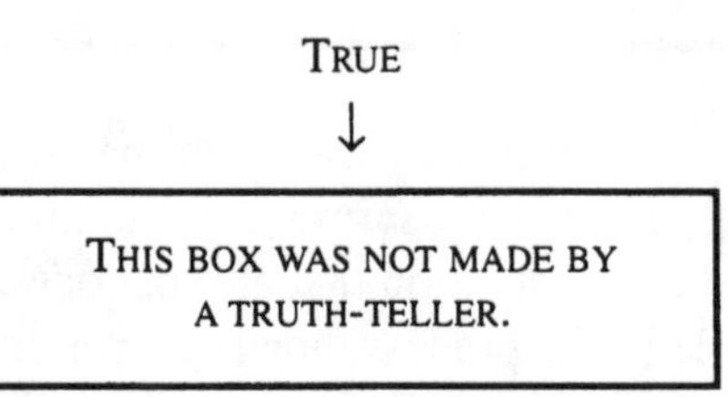
TRUE
↓
THIS BOX WAS NOT MADE BY
A TRUTH-TELLER.

But a liar would not make such a truthful statement. So, it is not possible to determine who made the box!

82 **Answer:** *Only $x = 0$ works, because that is the only value x can have in the equation.*

Solution: Solving the equation $x + y = y$, you'll get $x = 0$. Any other value assigned to x makes the equation impossible. There is no requirement that an equation is true for all possible values of its variables. Indeed, the object of solving an equation is to find the values that make it true.

83 **Answer:** *The truth value of the sentence cannot be determined.*

Solution: If you assume that the sentence is true, then it is true, or, to put it in the actual words of the sentence, it is "*not false*." Now, another way to say "*not false*," is to say that it is "*true*." So, let's rephrase the sentence (*This sentence is not false*) with its equivalent: *This sentence is true.* If you now assume that it is false, then it is, clearly, false. So, it is both true and false, no matter which way you look at it.

84 **Answer:** *It is impossible to make anything of what the sister said.*

Solution: If the sister never tells her brother the truth, then what she said yesterday was false. What did she say? "*I have never told you the truth.*" But wait! That is the truth! So, she didn't lie to her brother! Clearly, it is impossible to make anything of what she said yesterday, logically speaking, of course! From a different angle, she may have been playing a prank on her brother!

85 **Answer:** *Of course you have!*

Solution: Let's say that I am telling the truth. Then my statement—"*I'm lying when I say that you have not become a great puzzle solver*"—is true. Therefore,

I am lying when I say that you have *not* become a great puzzle solver. It appears, therefore, that you have in fact become a great puzzle solver.

Now let's say that I am lying. Then my statement is false. So, I will correct my statement to make it true: "*I'm not lying when I say that you have become a great puzzle solver*." So, it appears, again, that you have indeed become a great puzzle solver—no matter how you look at it.

Actually, this last puzzle was given to you purely in jest. Only you yourself will know how good a puzzle solver you have become.

11 A Puzzle IQ Test

This book has provided you with broad guidelines and analytical tools for solving various types of puzzles (puzzles in deductive logic, puzzles in truth logic, etc.). Its aim has been to help you enter into the channels of thought along which successful puzzle-solving moves. The following *Puzzle IQ Test* will give you an opportunity to find out how good you have become at puzzle-solving.

Before You Get Started

Above all else, success at puzzle-solving requires that you know how to analyze new puzzle conditions, data, facts, and so on, on the basis of what you already know. If you have forgotten the basic principles that apply to the solution of any one of the test puzzles, go back to, and review, the chapter that deals with the genre to which the puzzle belongs.

Before attempting to solve any specific test puzzle, ask yourself the following questions:

- To what genre does it belong? Is it a puzzle in deductive logic, in truth logic, and so on?
- Does it contain any deceptive statements or word tricks?
- Can the puzzle be reduced to something simpler?
- Can I compare it with any other puzzle I have already solved and whose principles I have grasped?
- What systematic approach (diagram, chart, etc.) can be employed?
- Are there any aspects of the puzzle that are irrelevant—that is, are they just there for stylistic embellishment or to confuse or misdirect me?

Good luck!

□□■ Puzzle IQ Test

1 Someone robbed a bank yesterday. Four suspects were rounded up and interrogated. One of them was indeed the robber. Here's what they said under severe questioning by the police:

Al: *Don did it.*
Don: *Tom did it.*
Guy: *I didn't do it.*
Tom: *Don lied when he said that I did it.*

Only one of these four statements turned out to be true. Who was the guilty man?

2 There are six oranges in a box. Without cutting up any of the oranges, Jessica divided the six oranges among six boys, and yet one orange was left in the box. How did she do this?

3 Bill, Paul, Ron, and Sam are musicians. One is a drummer, one a pianist, one a singer, and one a violinist, though not necessarily respectively.

1. Bill and Ron were in the audience when the singer performed as Mozart's Don Giovanni.
2. Both Paul and the violinist have attended the concerts of their friend, the pianist.
3. The violinist often performs with Sam and Bill.
4. Bill and Ron have never attended the pianist's concerts together.

What is each man's musical field?

4 There are 12 coins on a table arranged to form six equal squares:

COINS

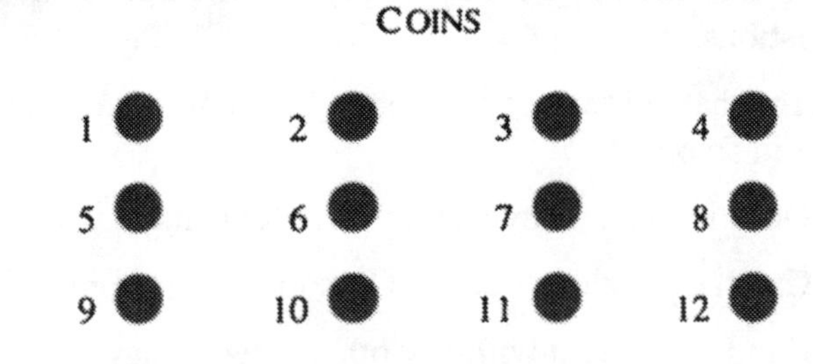

The six squares are as follows:

SQUARE 1 = 1-2-5-6

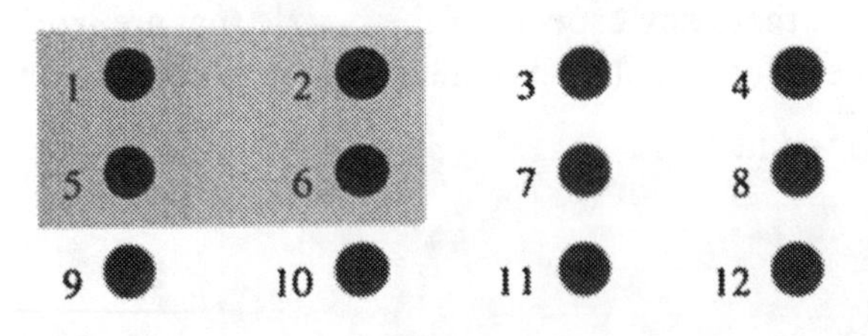

SQUARE 2 = 2-3-6-7

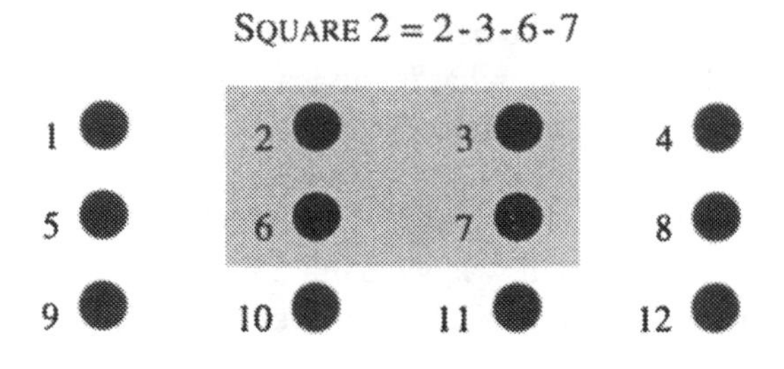

SQUARE 3 = 3-4-7-8

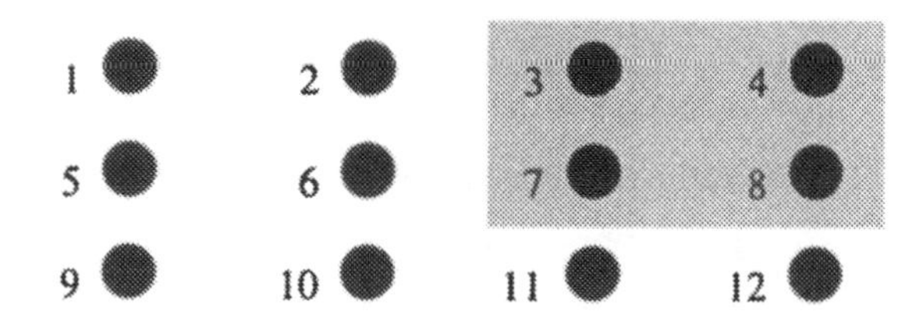

SQUARE 4 = 5-6-9-10

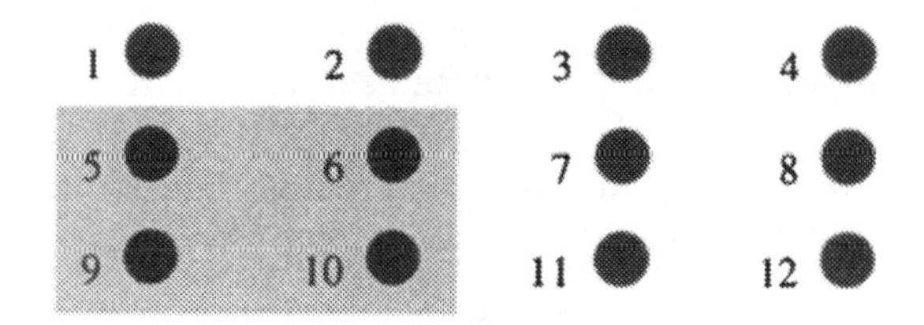

SQUARE 5 = 6-7-10-11

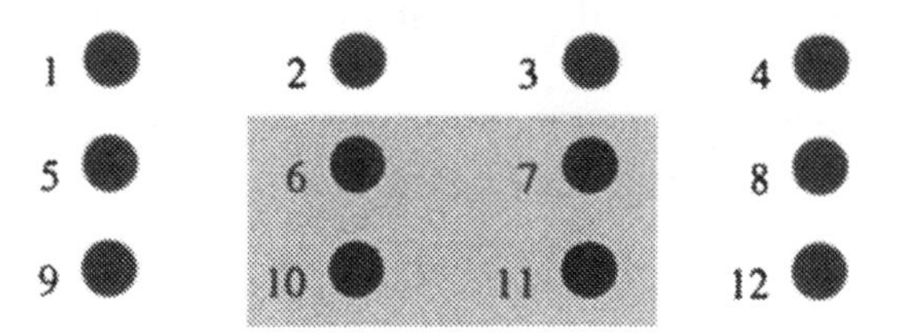

SQUARE 6 = 7-8-11-12

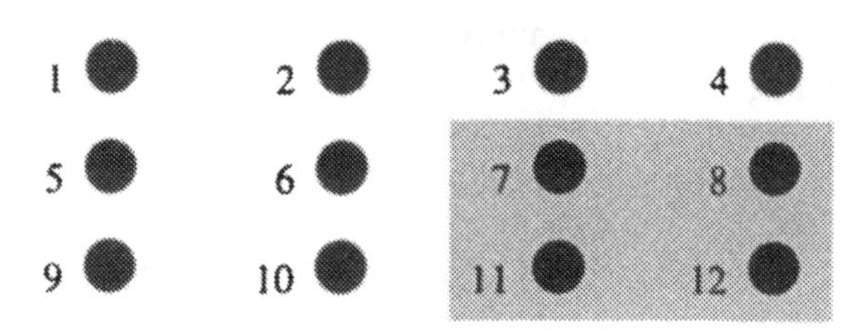

Can you remove just three coins to leave just three equal squares?

5 In a box there are 100 balls: 20 white, 20 black, 20 green, 20 blue, and 20 yellow. With a blindfold on, what is the least number you must draw out in order to get a pair of balls that matches?

6 Barb, Cam, Ines, Dick, and Mike were at a new yuppie café that opened just a few days ago. Each was wearing a different color T-shirt (black, blue, green, orange, or red) and each one ordered a different beverage (coffee, cola, hot chocolate, milk, or tea).

1. The person with the red T-shirt ordered cola.
2. Cam had hot chocolate.
3. Dick wore the blue T-shirt.
4. Mike did not order milk.
5. The person who ordered milk wore the orange T-shirt.
6. Barb wore the green T-shirt. She did not order coffee.

What color T-shirt did each person wear, and what beverage did each one order?

7 A grocery store has candy that sells at 45¢ per pound and nuts that sell at 30¢ per pound. A customer buys a mixture of 50 pounds of the two. The mixture sells for 40¢ per pound. How many pounds of each will be in the mixture?

8 What famous English saying is hidden in the following coded message? Note that 1 = T, 2 = I, and 5 = F.

1 2 3 4 5 6 2 4 7

9 A certain number is three less than twice another number. The two numbers added together equal 51. What are the numbers?

10 The following coded message refers to a saying concerning a day of the week. If 1 = N and 2 = E, what saying is it?

1 2 3 2 4 5 1 6 7 1 8 9 10

11 Alexander is three times as old as Jessica. In eight years, the sum of their ages will be 28 years. How old is each one?

"*All the people in my village are liars.*" Did this particular village member speak the truth?

13 Can you decipher the following alphametic?

```
    E  V  E
  + E  V  E
  ---------
    V  I  V
```

14 Bill can do a piece of work in 8 days, Mary in 10. How long will it take for them to do it working together?

15 Jack works every fourth day at the U-Save Department Store as a part-time sales clerk. Alicia also works there, but only on Saturdays. The store stays open seven days per week. This week, Jack worked on Monday, October 1. On what date will the two be working together?

16 Fiona is as old as the combined ages of her two brothers, Hank and Elway. Elway is two years older than Hank. Last year the combined ages of the three was three-fourths of their present combined ages. How old is each at present?

17 A farmer wants to fence his triangular chicken pen, which has one of its three equal sides against a complete side of his square barn. If the area of the barn is 100 square feet, how much fencing will he need?

18 Yesterday my best friend, Angelica, withdrew $50 from her account as follows:

$20	leaving	$30
$15	leaving	$15
$ 9	leaving	$ 6
$ 6	leaving	$ 0
$50		$ 51

Where did the extra dollar come from?

19 I have 15 billiard balls, one of which weighs less than the other 14. Otherwise, they all look the same. How can I identify the one that weighs less on a balance scale with no more than three weighings?

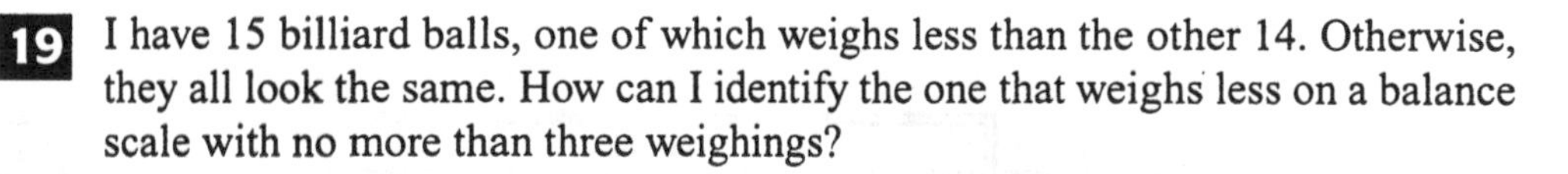

20 Maria has two dimes. If $\frac{4}{5}$ of what Maria has equals $\frac{8}{9}$ of what Betty has, how much money does Betty have?

Answers and Solutions

At this point, you should not need to have the solutions fully explained to you. Consequently, each solution will: (1) tell you what genre of puzzle it belongs to, and (2) refer you to a specific puzzle(s) in a previous chapter you can consult to get insights on how to solve it, and (3) go quickly through the main steps leading to a solution.

1 **Genre:** Truth logic [Chapter 2: puzzles 6, 10]

Answer: *Guy.*

Solution: Start by setting up a truth chart:

	STATEMENT	TRUTH VALUE
AL	DON DID IT	
DON	TOM DID IT	
GUY	I DIDN'T DO IT	
TOM	DON LIED WHEN HE SAID THAT I DID IT	

Since there is only one true statement, and since Don's statement—"*Tom did it*"—contradicts Tom's—"*Don lied when he said that I did it*"—then one of them is true and the other false. Assume that Don told the truth. His statement, therefore, has the only T-value:

	STATEMENT	TRUTH VALUE
AL	DON DID IT	F
DON	TOM DID IT	T
GUY	I DIDN'T DO IT	F
TOM	DON LIED WHEN HE SAID THAT I DID IT	F

Logically, if Don's statement is true, then Tom is the guilty party. But this leads to a contradiction. Consider Guy's statement—"*I didn't do it.*" Since it has an F-value, then it implies that he actually did it. But, then, there are two culprits—Tom and Guy. So, the original assumption—that Don's statement is true and Tom's false—can be rejected. But, in the process, you have learned with certainty that Don's statement is false and that Tom's statement is the only true one:

	STATEMENT	TRUTH VALUE
AL	DON DID IT	F
DON	TOM DID IT	F
GUY	I DIDN'T DO IT	F
TOM	DON LIED WHEN HE SAID THAT I DID IT	T

Now, consider each statement in the light of its truth-value. Al's statement that Don did it is false. So, Don did not do it. Don's statement that Tom did it is also false. So, Tom, too, did not do it. Tom's statement that Don lied when he said that he did it can now be seen to be true—as its T-value indicates. Guy's statement that he didn't do it is false. This means that Guy is the guilty man.

2 **Genre:** Trick logic [Chapter 3: puzzles 11, 12, 14, 15, 17]

Answer: *Jessica left one orange in the box, giving the box and its contents to one of the boys.*

Solution: The puzzle does not say that all the oranges were taken out of the box. So, obviously, Jessica left one of the oranges in the box, instead of taking it out, and gave it (along with the box) to one of the boys.

3 **Genre:** Deductive logic [Chapter 1: puzzles 1, 2, 3, 4]

Answer: *Bill is the drummer, Paul the singer, Ron the violinist, and Sam the pianist.*

Solution: Start by setting up an appropriate cell chart:

	DRUMMER	PIANIST	SINGER	VIOLINIST
BILL				
PAUL				
RON				
SAM				

From statement 1 it can be established that neither Bill nor Ron is the singer; from statement 2 that Paul is neither the violinist nor the pianist; and from statement 3 that neither Sam nor Bill is the violinist:

	DRUMMER	PIANIST	SINGER	VIOLINIST
BILL			×	×
PAUL		×		×
RON			×	
SAM				×

The chart now reveals that Ron is the violinist, because the only cell left under violinist is opposite Ron:

	DRUMMER	PIANIST	SINGER	VIOLINIST
BILL			×	×
PAUL		×		×
RON	×	×	×	●
SAM				×

Statement 2 tells you that both Paul and Ron (the violinist) have attended the concerts of their friend, the pianist. From statement 4 you can deduce that Bill is not the pianist.

	DRUMMER	PIANIST	SINGER	VIOLINIST
BILL		×	×	×
PAUL		×		×
RON	×	×	×	●
SAM				×

This leaves Bill as the drummer, because the only cell left opposite Bill is under drummer—and Sam as the pianist, because the only cell left under pianist is opposite Sam:

	DRUMMER	PIANIST	SINGER	VIOLINIST
BILL	●	×	×	×
PAUL	×	×		×
RON	×	×	×	●
SAM	×	●	×	×

This leaves Paul as the singer:

	DRUMMER	PIANIST	SINGER	VIOLINIST
BILL	●	×	×	×
PAUL	×	×	●	×
RON	×	×	×	●
SAM	×	●	×	×

In conclusion, Bill is the drummer, Paul the singer, Ron the violinist, and Sam the pianist.

4 **Genre:** Geometrical logic [Chapter 7: puzzles 56, 57]
Answer: *Remove coins 4, 9, and 10.*
Solution:

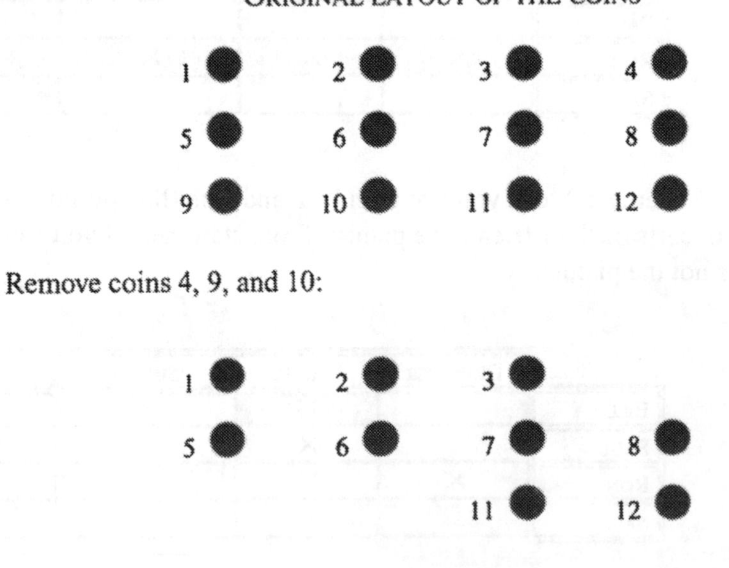

The three squares are:

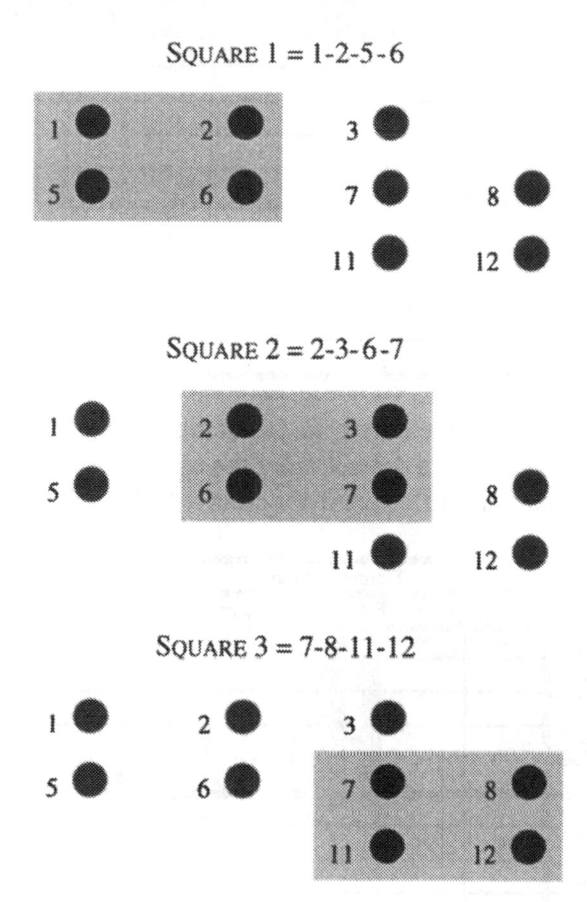

5 **Genre:** Combinatory logic [Chapter 6: puzzles 45, 48]

Answer: *Six.*

Solution: In this case the worst-case scenario involves drawing 5 balls all of different colors—1 white, 1 black, 1 green, 1 blue, and 1 yellow. Then, the *sixth* ball you take out will produce a match—a white ball produces a pair of white balls, a black ball produces a pair of black balls, a green ball produces a pair of green balls, a blue ball produces a pair of blue balls, or a yellow ball produces a pair of yellow balls.

6 **Genre:** Deductive logic [Chapter 1: puzzle 5]

Answer: *Barb wore the green T-shirt and ordered tea; Cam wore the black T-shirt and ordered hot chocolate; Ines wore the orange T-shirt and ordered milk; Dick wore the blue T-shirt and ordered coffee; and Mike wore the red T-shirt and ordered cola.*

Solution: Set up an appropriate cell chart. From statement 1, it can be established that the person who wore the red T-shirt ordered the cola:

		T-shirt colors					Beverages				
		Black	Blue	Green	Orange	Red	Coffee	Cola	Choc.	Milk	Tea
Names	Barb										
	Cam										
	Ines										
	Dick										
	Mike										
Beverages	Coffee					×					
	Cola	×	×	×	×	●					
	Choc.					×					
	Milk					×					
	Tea					×					

From statement 6, it can be established that: (1) Barb wore the green T-shirt; (2) she therefore did not order the cola (because the one who wore a red T-shirt did that); and (3) she did not order a coffee:

		T-shirt colors					Beverages				
		Black	Blue	Green	Orange	Red	Coffee	Cola	Choc.	Milk	Tea
Names	Barb	×	×	●	×	×	×	×			
	Cam			×							
	Ines			×							
	Dick			×							
	Mike			×							
Beverages	Coffee			×		×					
	Cola	×	×	×	×	●					
	Choc.					×					
	Milk					×					
	Tea					×					

From statement 2, it can be established that: (1) Cam ordered the hot chocolate; (2) the green T-shirt wearer, who is not Cam, therefore did not order hot chocolate:

		T-shirt colors					Beverages				
		Black	Blue	Green	Orange	Red	Coffee	Cola	Choc.	Milk	Tea
Names	Barb	×	×	●	×	×	×	×	×		
	Cam			×			×	×	●	×	×
	Ines			×					×		
	Dick			×					×		
	Mike			×					×		
Beverages	Coffee			×		×					
	Cola	×	×	×	×	●					
	Choc.			×		×					
	Milk					×					
	Tea					×					

Statement 3 tells you that Dick wore the blue T-shirt:

		T-SHIRT COLORS					BEVERAGES				
		BLACK	BLUE	GREEN	ORANGE	RED	COFFEE	COLA	CHOC.	MILK	TEA
NAMES	BARB	×	×	●	×	×	×	×	×		
	CAM		×	×			×	×	●	×	×
	INES		×	×					×		
	DICK	×	●	×	×	×			×		
	MIKE		×	×					×		
BEVERAGES	COFFEE			×		×					
	COLA	×	×	×	×	●					
	CHOC.			×		×					
	MILK					×					
	TEA					×					

The chart shows that Dick did not order hot chocolate. Since Dick is the blue T-shirt wearer, the blue T-shirt possibility opposite hot chocolate can now be eliminated:

		T-SHIRT COLORS					BEVERAGES				
		BLACK	BLUE	GREEN	ORANGE	RED	COFFEE	COLA	CHOC.	MILK	TEA
NAMES	BARB	×	×	●	×	×	×	×	×		
	CAM		×	×			×	×	●	×	×
	INES		×	×					×		
	DICK	×	●	×	×	×			×		
	MIKE		×	×					×		
BEVERAGES	COFFEE			×		×					
	COLA	×	×	×	×	●					
	CHOC.		×	×		×					
	MILK					×					
	TEA					×					

Statement 5 informs you that the person who ordered milk wore the orange T-shirt:

		T-SHIRT COLORS					BEVERAGES				
		BLACK	BLUE	GREEN	ORANGE	RED	COFFEE	COLA	CHOC.	MILK	TEA
NAMES	BARB	×	×	●	×	×	×	×	×		
	CAM		×	×			×	×	●	×	×
	INES		×	×					×		
	DICK	×	●	×	×	×			×		
	MIKE		×	×					×		
BEVERAGES	COFFEE			×	×	×					
	COLA	×	×	×	×	●					
	CHOC.		×	×	×	×					
	MILK	×	×	×	●	×					
	TEA				×	×					

The lower set now reveals that: (1) the black T-shirt wearer ordered hot chocolate; (2) the green T-shirt-wearer ordered tea:

		T-SHIRT COLORS					BEVERAGES				
		BLACK	BLUE	GREEN	ORANGE	RED	COFFEE	COLA	CHOC.	MILK	TEA
NAMES	BARB	×	×	●	×	×	×	×	×		
	CAM		×	×			×	×	●	×	×
	INES		×	×					×		
	DICK	×	●	×	×	×			×		
	MIKE		×	×					×		
BEVERAGES	COFFEE	×		×	×	×					
	COLA	×	×	×	×	●					
	CHOC.	●	×	×	×	×					
	MILK	×	×	×	●	×					
	TEA	×	×	●	×	×					

That leaves the blue T-shirt wearer as the one who ordered coffee:

		T-SHIRT COLORS					BEVERAGES				
		BLACK	BLUE	GREEN	ORANGE	RED	COFFEE	COLA	CHOC.	MILK	TEA
NAMES	BARB	×	×	●	×	×	×	×	×		
	CAM		×	×			×	×	●	×	×
	INES		×	×					×		
	DICK	×	●	×	×	×			×		
	MIKE		×	×					×		
BEVERAGES	COFFEE	×	●	×	×	×					
	COLA	×	×	×	×	●					
	CHOC.	●	×	×	×	×					
	MILK	×	×	×	●	×					
	TEA	×	×	●	×	×					

The chart now allows you to make the following correlations: (1) Since Barb is the green T-shirt wearer, and the green T-shirt wearer ordered tea, then Barb is the one who ordered tea. (2) Since the person who ordered the coffee wore a blue T-shirt, and the blue T-shirt wearer is Dick, then Dick is the one who ordered coffee:

		T-SHIRT COLORS					BEVERAGES				
		BLACK	BLUE	GREEN	ORANGE	RED	COFFEE	COLA	CHOC.	MILK	TEA
NAMES	BARB	×	×	●	×	×	×	×	×	×	●
	CAM		×	×			×	×	●	×	×
	INES		×	×			×		×		×
	DICK	×	●	×	×	×	●	×	×	×	×
	MIKE		×	×			×		×		×
BEVERAGES	COFFEE	×	●	×	×	×					
	COLA	×	×	×	×	●					
	CHOC.	●	×	×	×	×					
	MILK	×	×	×	●	×					
	TEA	×	×	●	×	×					

Statement 4 tells you that Mike did not order milk:

		T-SHIRT COLORS					BEVERAGES				
		BLACK	BLUE	GREEN	ORANGE	RED	COFFEE	COLA	CHOC.	MILK	TEA
NAMES	BARB	×	×	●	×	×	×	×	×	×	●
	CAM		×	×			×	×	●	×	×
	INES		×	×			×		×		×
	DICK	×	●	×	×	×	●	×	×	×	×
	MIKE		×	×			×		×	×	×
BEVERAGES	COFFEE	×	●	×	×	×					
	COLA	×	×	×	×	●					
	CHOC.	●	×	×	×	×					
	MILK	×	×	×	●	×					
	TEA	×	×	●	×	×					

The right-hand set now shows that Ines is the one who ordered milk. Thus, since the orange T-shirt wearer ordered milk, then Ines was the orange T-shirt wearer:

		T-SHIRT COLORS					BEVERAGES				
		BLACK	BLUE	GREEN	ORANGE	RED	COFFEE	COLA	CHOC.	MILK	TEA
NAMES	BARB	×	×	●	×	×	×	×	×	×	●
	CAM		×	×	×		×	×	●	×	×
	INES	×	×	×	●	×	×	×	×	●	×
	DICK	×	●	×	×	×	●	×	×	×	×
	MIKE		×	×	×		×		×	×	×
BEVERAGES	COFFEE	×	●	×	×	×					
	COLA	×	×	×	×	●					
	CHOC.	●	×	×	×	×					
	MILK	×	×	×	●	×					
	TEA	×	×	●	×	×					

Finally, the right-hand set shows that Mike was the one who ordered cola. Hence, Mike wore the red T-shirt, since the one who wore this color of T-shirt was the one who ordered cola. This leaves Cam as the one who wore the black T-shirt and who, therefore, ordered hot chocolate:

		T-SHIRT COLORS					BEVERAGES				
		BLACK	BLUE	GREEN	ORANGE	RED	COFFEE	COLA	CHOC.	MILK	TEA
NAMES	BARB	×	×	●	×	×	×	×	×	×	●
	CAM	●	×	×	×	×	×	×	●	×	×
	INES	×	×	×	●	×	×	×	×	●	×
	DICK	×	●	×	×	×	●	×	×	×	×
	MIKE	×	×	×	×	●	×	●	×	×	×
BEVERAGES	COFFEE	×	●	×	×	×					
	COLA	×	×	×	×	●					
	CHOC.	●	×	×	×	×					
	MILK	×	×	×	●	×					
	TEA	×	×	●	×	×					

In summary: Barb wore the green T-shirt and ordered tea; Cam wore the black T-shirt and ordered hot chocolate; Ines wore the orange T-shirt and ordered milk; Dick wore the blue T-shirt and ordered coffee; and Mike wore the red T-shirt and ordered cola.

7 **Genre:** Algebraic logic [Chapter 5: puzzles 40, 44]

Answer: *33⅓ pounds of candy, 16⅔ pounds of nuts.*

Solution: Start by letting x stand for the number of pounds of candy in the mixture. Therefore, the number of pounds of nuts in the mixture is $50 - x$.

The candy costs \$.45 per pound. So, x pounds cost $.45x$. The nuts cost \$.30 per pound. So, $50 - x$ nuts cost $.30(50 - x)$. The total value of the candy and nuts together is, therefore: $.45x + .30(50 - x)$.

You are told that the mixture costs \$.40 per pound. So, 50 pounds cost $\$.40 \times 50 = \$20$. This is, of course, the cost of the candy and nuts together:

$$.45x + .30(50 - x) = 20$$

Solving for x, you'll get: $x = 33\frac{1}{3}$ pounds. So, there will be 33⅓ pounds of candy in the mixture and 16⅔ (= 50 – 33⅓) pounds of nuts in it.

Check this out: 33⅓ pounds of candy at \$.45 per pound is \$15; 16⅔ pounds of nuts at \$.30 per pound is \$5. Together, they cost \$15 + \$5 = \$20.

8 **Genre:** Code logic [Chapter 8: puzzles 58, 59, 60, 61]

Answer: *Time flies.*

Solution: Start by making the substitutions 1 = T, 2 = I, and 5 = F:

1	2	3	4		5	6	2	4	7
↑	↑				↑		↑		
T	I				F		I		

Consider the possibilities for the first word: *Tick, Tide, Ties, Tile, Time, Tire.* Since you are deciphering a saying, it would seem that *time* is the most likely candidate (having been the topic of much proverbializing throughout history). This would mean that 3 = M and 4 = E:

1	2	3	4		5	6	2	4	7
↑	↑	↑	↑		↑		↑	↑	
T	I	M	E		F		I	E	

The solution is, clearly, *Time flies*:

1	2	3	4		5	6	2	4	7
↑	↑	↑	↑		↑	↑	↑	↑	↑
T	I	M	E		F	L	I	E	S

9 **Genre:** Algebraic logic [Chapter 5: puzzle 36]

Answer: *18, 33.*

Solution: Start by letting x represent one of the two numbers. Then, the other number is $51 - x$, given that they add up to 51. One of the numbers, say $51 - x$, is three less than twice x:

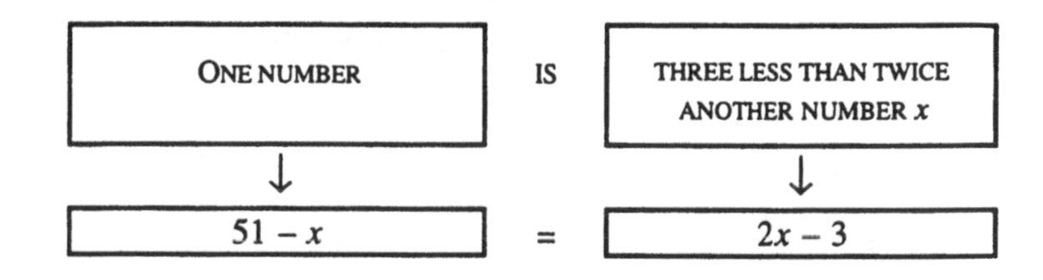

Solving for x, you'll get: $x = 18$. So, the numbers are 18 and 33 (= 51 − 18). Check this. Twice 18 is 36; and three less than that is 33. This is, of course, the other number.

10 **Genre:** Code logic [Chapter 8: puzzles 58, 59, 60, 61]

Answer: *Never on Sunday.*

Solution: Substitute 1 = N and 2 = E in the message:

1	2	3	2	4		5	1		6	7	1	8	9	10
↑	↑		↑				↑				↑			
N	E		E				N				N			

The message refers to a day of the week. That day cannot be the first or second word, as you can see. So, it is the third word. Since its third letter is N, it is either Monday or Sunday. In any case, the last three letters are 8 = D, 9 = A, 10 = Y:

1	2	3	2	4		5	1		6	7	1	8	9	10
↑	↑		↑				↑				↑	↑	↑	↑
N	E		E				N				N	D	A	Y

Consider the first word. Try out a few letters for 3 and 4, and you will soon come to the conclusion that the word can only be *never.* Now, the rest of the puzzle is easily solved. The saying is, of course, *Never on Sunday*:

1	2	3	2	4		5	1		6	7	1	8	9	10
↑	↑	↑	↑	↑		↑	↑		↑	↑	↑	↑	↑	↑
N	E	V	E	R		O	N		S	U	N	D	A	Y

11 **Genre:** Algebraic logic [Chapter 5: puzzles 39, 41]
Answer: *Jessica is 3 years old and Alexander is 9 years old.*
Solution: Let x stand for Jessica's age. Then, Alexander is $3x$ years old—because he is three times as old as Jessica.

How old will Jessica be 8 years from now? She will, of course, be $x + 8$. And Alexander will be $3x + 8$. Together, their ages will add up to 28:

$$x + 8 + 3x + 8 = 28$$

Solving for x, you'll get: $x = 3$. So, Jessica is 3 years old and Alexander is 9 years old. Check this out. In eight years, Jessica will be 11 and Alexander will be 17. Together, their ages add up to $11 + 17 = 28$.

12 **Genre:** Paradox logic [Chapter 10: puzzles 74, 75, 76, 83, 84, 85]
Answer: *The statement is contradictory.*
Solution: The person who made the statement is himself a village member. So, he is a liar, and his statement is false. What is his statement? *All the people in my village are liars.* But, as you can see, the statement is true! The people in his village are indeed all liars. So, he is not a liar after all. He told it like it is. But, then, he is both a liar and not a liar!

13 **Genre:** Code logic [Chapter 8: puzzles 62, 63, 64, 65, 66]
Answer: *121 + 121 = 242* or *242 + 242 = 484.*
Solution: Since there is no carryover, **E** is less than 5. Start by letting **E** = 4:

	4	V	4
+	4	V	4
	V	I	V

This makes **V** = 8:

	4	8	4
+	4	8	4
	8	I	8

But, then, this creates a carryover in the middle column: 8 + 8 = 16. So, try **E** = 3:

	3	V	3
+	3	V	3
	V	I	V

This makes **V** = 6:

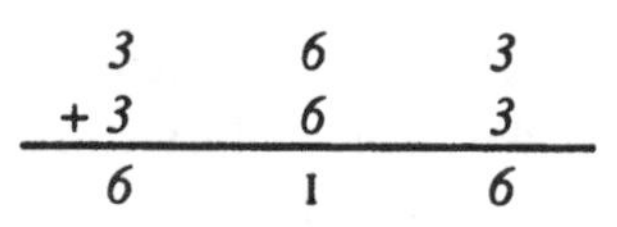

But, once again, the middle column produces a carryover: 6 + 6 = 12. Try **E** = 2:

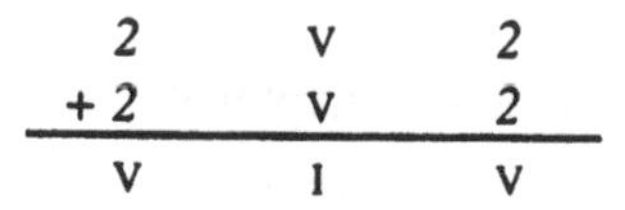

This makes **V** = 4:

2	4	2
+ 2	4	2
4	I	4

In the middle column, 4 + 4 = 8. So, **I** = 8. This is the solution:

2	4	2
+ 2	4	2
4	8	4

Check and see if there is a second solution, by substituting **E** = 1:

1	V	1
+ 1	V	1
V	I	V

This makes **V** = 2:

1	2	1
+ 1	2	1
2	I	2

This means that **I** = 4. This is, in fact, a second solution to the puzzle:

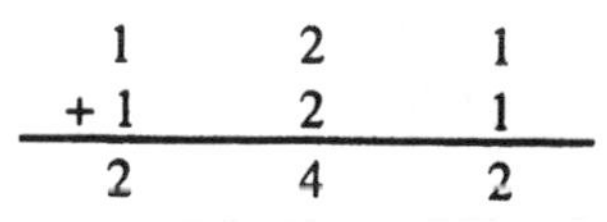

14 **Genre:** Time logic [Chapter 9: puzzles 69, 73]

Answer: *$4\frac{4}{9}$ days.*

Solution: Since Bill can complete a piece of work in 8 days, he can complete $\frac{1}{8}$ of the job per day. Since Mary can complete a piece of work in 10 days, she can complete $\frac{1}{10}$ of it per day.

Now, let x stand for the number of days Bill and Mary need to complete the job working together. In x days, Bill will complete $\frac{1}{8}x$ and Mary $\frac{1}{10}x$ of the whole job: $\frac{1}{8}x + \frac{1}{10}x = 1$. Solving for x, you'll get $x = 4\frac{4}{9}$. So, it will take $4\frac{4}{9}$ days for Bill and Mary to complete the piece of work together.

15 **Genre:** Time logic [Chapter 9: puzzle 72]

Answer: *Saturday, October 13.*

Solution: Set up a time chart for the week of October 1, showing that Jack (= J) works every fourth day and Alicia (= A) on Saturday:

WEEK OF OCTOBER 1

CLERK	MONDAY, OCT. 1	TUESDAY, OCT. 2	WEDNESDAY, OCT. 3	THURSDAY, OCT. 4	FRIDAY, OCT. 5	SATURDAY, OCT. 6	SUNDAY, OCT. 7
JACK	J			J			J
ALICIA						A	

Since they will apparently not be working together during that week, set up a chart for the week after:

WEEK OF OCTOBER 8

CLERK	MONDAY, OCT. 8	TUESDAY, OCT. 9	WEDNESDAY, OCT. 10	THURSDAY, OCT. 11	FRIDAY, OCT. 12	SATURDAY, OCT. 13	SUNDAY, OCT. 14
JACK			J			J	
ALICIA						A	

This chart shows that the two will be working together on Saturday, October 13.

16 **Genre:** Algebraic logic [Chapter 5: puzzles 39, 41]

Answer: *Hank is 2 years old, Elway 4 years old, and Fiona 6 years old.*

Solution: Let x stand for Hank's present age. Therefore, Elway, who is two years older than Hank, is $x + 2$ years old; and Fiona, who is as old as her two brothers combined, is $x + x + 2$, or $2x + 2$, years old. Their combined age this year is, therefore: $x + (x + 2) + (2x + 2) = 4x + 4$.

What was the age of each one last year? It was 1 year less than his or her present age:

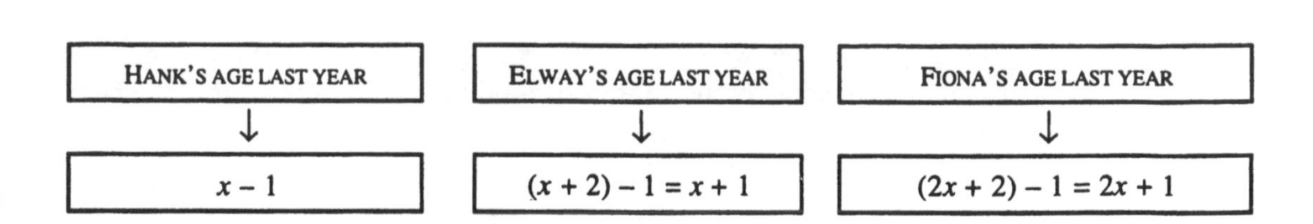

Their combined age last year was, therefore: $(x - 1) + (x + 1) + (2x + 1) = 4x + 1$. This equals three-fourths of their combined age at present: $4x + 1 = \frac{3}{4}(4x + 4)$. Solving for x, you'll get: $x = 2$. So, Hank is 2 years old, Elway 4 years old, and Fiona 6 years old. Check this. Last year Hank was 1, Elway 3, and Fiona 5. The sum of their ages was $1 + 3 + 5 = 9$. The sum of their present ages is $2 + 4 + 6 = 12$. And 9 is indeed three-fourths of 12.

17 **Genre:** Geometrical logic [Chapter 7: puzzles 54, 55]

Answer: *20 feet.*

Solution: Draw a figure of the square barn with the triangular pen against one of its sides. The sides of the barn and pen are all equal. This is because the barn is a square and because one of the sides of the equilateral triangular pen is equal to one of the sides of the barn. Let x stand for the length of a side in feet:

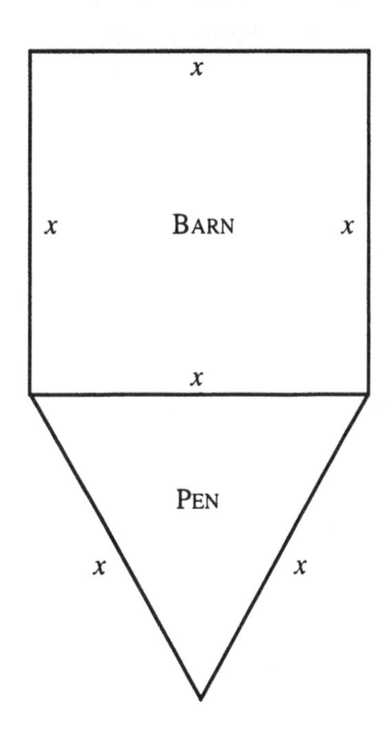

The area of the barn, $x \times x$, is 100 square feet:

$$x \times x = 100$$

$$x^2 = 100$$

$$x = 10$$

So, the lengths of the two sides of the rectangular pen are 10 feet each. The farmer will need fencing for just those sides—the ones jutting out from the barn. So, he will need 20 feet of fencing.

18 **Genre:** Trick logic [Chapter 3: puzzles 15, 18, 19]

Answer: *In addition, the sum of leftovers (the right column figures) does not equal the sum of the amounts subtracted (the left column figures).*

Solution: The column on the left side contains the actual dollars that the friend took out. These add up, as they should, to $50. But the figures in the column on the right side are *leftovers*: that is, when Angelica took out $20, she had $50 – $20 = $30 left over; when she took out the next $15, she had $30 – $15 = $15 left over; and so on. There is no reason for these *leftovers* to add up to $50, is there? In arithmetical language, the sum of the leftovers (the right column figures) does not equal the sum of the amounts subtracted (the left column figures).

19 **Genre:** Arithmetical logic [Chapter 4: puzzle 25]

Answer: *It can be done by dividing the balls into two equal batches of 7 balls (first weighing), 3 balls (second weighing), and 1 ball (third weighing), and putting one ball aside each time.*

Solution: First, put aside one of the fifteen balls, leaving two equal batches of seven balls each. Put these two batches on a balance—seven on the left pan and seven on the right pan.

- If the balls balance, then the ball that was put aside is, obviously, the one that weighs less. So, in only one weighing you would have identified the culprit ball.
- If the balls do not balance, then you know that the culprit ball is on the pan that went up. So, discard the other seven balls, and proceed to a second weighing.
- Put aside one of the seven suspect balls, leaving two equal batches of three balls each. Put these on the balance—three on the left pan and three on the right pan. If the balls balance, then the ball that was put aside is the one that weighs less. In such an event, it has taken just two weighings to identify the culprit ball. If they do not, then the culprit ball is on the pan that went up. Discard the other three balls and proceed to a third weighing.
- Put aside one of the three suspect balls, leaving two balls. Put one on the left pan and one on the right pan. If the balls balance then the ball that was put aside is the culprit ball. If they do not, then the ball that weighs less is the one on the pan that goes up.

20 **Genre:** Arithmetical logic [Chapter 4: puzzle 31]; can also be solved as a puzzle in algebraic logic [Chapter 5: puzzle 40]

Answer: *18¢.*

Solution: Maria has two dimes, or 20¢. You are told that $\frac{4}{5}$ of what Maria has equals $\frac{8}{9}$ of what Betty has. So, how much is $\frac{4}{5}$ of what Maria has? It is $\frac{4}{5}$ of 20¢ = 16¢. This equals $\frac{8}{9}$ of what Betty has: $\frac{8}{9}$ of Betty's money = 16¢. Betty, thus, has 16¢ ÷ $\frac{8}{9}$ = 18¢.

Score

18–20 correct	=	*Excellent*
15–17 correct	=	*Very good*
11–14 correct	=	*Good*
8–10 correct	=	*Passing grade*
Below 8 correct	=	*Go over the entire book again for more practice*

About the Author

Marcel Danesi (Ph.D.) is Professor of Education and Communication Theory at the University of Toronto and at the Ontario Institute for Studies in Education. For over 25 years, he has been conducting research on how children and adults learn. He has developed educational materials designed to help learners of all ages and from different linguistic and cultural backgrounds acquire facility in solving problems in mathematics. He has published several books on the role of problem-solving in education, including *Puzzles and Games in Language Teaching* (1989) and *Problem-Solving and Second-Language Teaching* (1992).

Printed in the USA
CPSIA information can be obtained
at www.ICGtesting.com
JSHW052016140824
68134JS00027B/2510

9 780471 157250